Astounding Stories
Of Super-Science
Vol. 1

by

Ed. Harry Bates

Double 9
BOOKS

Astounding Stories Of Super-Science
Vol. 1
by Ed. Harry Bates

ISBN: 978-93-59320-39-7

Published by

DOUBLE 9 BOOKS

2/13-B, Ansari Road
Daryaganj, New Delhi – 110002
info@double9books.com
www.double9books.com
Tel. 011-40042856

ABOUT THE EDITOR

Ed. Harry Bates, whose full name was Harry Bates (1900-1981), was an American science fiction writer and editor who played a significant role in the early development of the genre. He is best known for his work as the founding editor of "Astounding Science Fiction" magazine (later known as "Analog Science Fiction and Fact"). Under Bates' editorial guidance, "Astounding Science Fiction" became one of the most influential and respected publications in the field of science fiction during the 1930s and 1940s. He had a keen eye for talent and published stories by some of the genre's most celebrated authors, including Isaac Asimov, A. E. van Vogt, and Robert A. Heinlein. It was during Bates' tenure that many groundbreaking and thought-provoking science fiction stories were first introduced to the public. While Bates was primarily known for his work as an editor, he also wrote several science fiction stories himself, contributing to the genre he helped shape. His most famous work as a writer is "Farewell to the Master, " which was later adapted into the classic science fiction film "The Day the Earth Stood Still" in 1951. Harry Bates' contributions to both the literary and editorial sides of science fiction have left a lasting impact on the genre.

CONTENTS

Introducing — Astounding Stories

What *are* "astounding" stories?

Well, if you lived in Europe in 1490, and someone told you the earth was round and moved around the sun—that would have been an "astounding" story.

Or if you lived in 1840, and were told that some day men a thousand miles apart would be able to talk to each other through a little wire—or without any wire at all—that would have been another.

Or if, in 1900, they predicted ocean-crossing airplanes and submarines, world-girdling Zeppelins, sixty-story buildings, radio, metal that can be made to resist gravity and float in the air—these would have been other "astounding" stories.

To-day, time has gone by, and all these things are commonplace. That is the only real difference between the astounding and the commonplace—Time.

To-morrow, more astounding things are going to happen. Your children—or their children—are going to take a trip to the moon. They will be able to render themselves invisible—a problem that has already been partly solved. They will be able to disintegrate their bodies in New York and reintegrate them in China—and in a matter of seconds.

Astounding? Indeed, yes.

Impossible? Well—television would have been impossible, almost unthinkable, ten years ago.

Now you will see the kind of magazine that it is our pleasure to offer you beginning with this, the first number of Astounding Stories.

It is a magazine whose stories will anticipate the super-scientific achievements of To-morrow—whose stories will not only be strictly accurate in their science but will be vividly, dramatically and thrillingly told.

Already we have secured stories by some of the finest writers of fantasy in the world—men such as Ray Cummings, Murray Leinster, Captain S. P. Meek, Harl Vincent, R. F. Starzl and Victor Rousseau.

So—order your next month's copy of Astounding Stories in advance!—*The Editor.*

THE BEETLE HORDE

A TWO-PART NOVEL

By *Victor Rousseau*

CHAPTER I
DODD'S DISCOVERY

Only two young explorers stand in the way of the mad Bram's horrible revenge—the releasing of his trillions of man-sized beetles upon an utterly defenseless world.

Out of the south the biplane came winging back toward the camp, a black speck against the dazzling white of the vast ice-fields that extended unbroken to the horizon on every side.

It came out of the south, and yet, a hundred miles further back along the course on which it flew, it could not have proceeded in any direction except northward. For a hundred miles south lay the south pole, the goal toward which the Travers Expeditions had been pressing for the better part of that year.

Not that they could not have reached it sooner. As a matter of fact, the pole had been crossed and re-crossed, according to the estimate of Tommy Travers, aviator, and nephew of the old millionaire who stood fairy uncle to the expedition. But one of the things that was being sought was the exact site of the pole. Not within a couple of miles or so, but within the fraction of an inch.

It had something to do with Einstein, and something to do with terrestrial magnetism, and the variations of the south magnetic pole, and the reason therefore, and something to do with parallaxes and the precession of the equinoxes and other things, this search for the pole's exact location. But all that was principally the affair of the astronomer of the party. Tommy Travers, who was now evidently on his way back, didn't give a whoop for Einstein, or any of the rest of the stuff. He had been enjoying himself after his fashion during a year of frostbites and hard rations, and he was beginning to anticipate the delights of the return to Broadway.

Captain Storm, in charge of the expedition, together with the five others of the advance camp, watched the plane maneuver up to the tents. She came down neatly on the smooth snow, skidded on her runners like an expert skater, and came to a stop almost immediately in front of the marquee.

Tommy Travers leaped out of the enclosed cockpit, which, shut off by glass from the cabin, was something like the front seat of a limousine.

"Well, Captain, we followed that break for a hundred miles, and there's no ground cleft, as you expected," he said. "But Jim Dodd and I picked up something, and Jim seems to have gone crazy."

Through the windows of the cabin, Jim Dodd, the young archaeologist of the party, could be seen apparently wrestling with something that looked like a suit of armor. By the time Captain Storm, Jimmy, and the other members of the party had reached the cabin door, Dodd had got it open and flung himself out backward, still hugging what he had found, and maneuvering so that he managed to fall on his back and sustain its weight.

"Say, what the—what—what's that?" gasped Storm.

Even the least scientific minded of the party gasped in amazement at what Dodd had. It resembled nothing so much as an enormous beetle. As a matter of fact, it was an insect, for it had the three sections that characterize this class, but it was merely the shell of one. Between four and five feet in height, when Dodd stood it on end, it could now be seen to consist of the hard exterior substance of some huge, unknown coleopter.

This substance, which was fully three inches thick over the thorax, looked as hard as plate armor.

"What is it?" gasped Storm again.

Tommy Travers made answer, for James Dodd was evidently incapable of speech, more from emotion than from the force with which he had landed backward in the snow.

"We found it at the pole, Captain," he said. "At least, pretty near where the pole ought to be. We ran into a current of warm air or something. The snow had melted in places, and there were patches of bare rock. This thing was lying in a hollow among them."

"If I didn't see it before my eyes, I'd think you crazy, Tommy," said Storm with some asperity. "What is it, a crab?"

"Crab be damned!" shouted Jim Dodd, suddenly recovering his faculties. "My God, Captain Storm, don't you know the difference between an insect and a crustacean? This is a fossil beetle. Don't you see the distinguishing mark of the coleoptera, those two elytra, or wing-covers, which meet in the median dorsal line? A beetle, but with the shell of a crustacean instead of mere chitin. That's what led you astray, I expect. God, what a tale we'll have to tell when we get back to New York! We'll drop everything else, and spend years, if need be, looking for other specimens."

"Like fun you will!" shouted Higby, the astronomer of the party. "Lemme tell you right here, Dodd, nobody outside the Museum of Natural History is going to care a damn about your old fossils. What we're going to do is to march straight to the true pole, and spend a year taking observations and parallaxes. If Einstein's brochure, in which he links up gravitation with magnetism, is correct—"

"Fossil beetles!" Jim Dodd burst out, ignoring the astronomer. "That means that in the Tertiary Era, probably, there existed forms of life in the antarctic continent that have never been found elsewhere. Imagine a world in which the insect reached a size proportionate to the great saurians, Captain Storm! I'll wager poor Bram discovered this. That's why he stayed behind when the Greystoke Expedition came within a hundred miles of the pole. I'll wager he's left a cairn somewhere with full details inside it. We've got to find it. We—"

But Jim Dodd, suddenly realizing that the rest of the party could hardly be said to share his enthusiasm in any marked degree, broke off and looked sulky.

"You say you found this thing pretty nearly upon the site of the true pole?" Captain Storm asked Tommy.

"Within five miles, I'd say, Captain. The fog was so bad that we couldn't get our directions very well."

"Well, then, there's going to be no difficulty," answered Storm. "If this fair weather lasts, we'll be at the pole in another week, and we'll start making our permanent camp. Plenty of opportunity for all you gentlemen. As for me, I'm merely a sailor, and I'm trying to be impartial.

"And please remember, gentlemen, that we're well into March now, and likely to have the first storms of autumn on us any day. So let's drop the argument and remember that we've got to pull together!"

Tommy Travers was the only skilled aviator of the expedition, which had brought two planes with it. It was a queer friendship that had sprung up between him and Jim Dodd. Tommy, the blasé ex-Harvard man, who was known along Broadway, and had never been able to settle down, seemed as different as possible from the spectacled, scholarly Dodd, ten years his senior, red-haired, irascible, and living, as Tommy put it, in the Age of Old Red Sandstone, instead of in the year 1930 A. D.

It was generally known—though the story had been officially denied—that there had been trouble in the Greystoke Expedition of three years before. Captain Greystoke had taken the brilliant, erratic Bram, of the Carnegie Archaeological Institute, with him, and Bram's history was a long record of trouble.

It was Bram who had exploded the faked neolithic finds at Mannheim, thereby earning the undying enmity of certain European savants, but brilliantly demolishing them when he smashed the so-called Mannheim stone pitcher (valued at a hundred thousand dollars) with a pocket-axe, and caustically inquired whether neolithic man used babbit metal rivets to fasten on his jug handles.

Bram's brilliant work in the investigation of the origin of the negrito Asiatic races had been awarded one of the Nobel prizes, and Bram had declined it in an insulting letter because he disapproved of the year's prize award for literature.

He had been a storm center for years, embittered by long opposition, when he joined the Greystoke Expedition for the purpose of investigating the marine fauna of the antarctic continent.

And it was known that his presence had nearly brought the Greystoke Expedition to the point of civil war. Rumor said he had been deliberately abandoned. His enemies hoped he had. The facts seemed to be, however, that in an outburst of temper he had walked out of camp in a furious snowstorm and perished. For days his body had been sought in vain.

Jimmy Dodd had run foul of Bram some years before, when Bram had published a criticism of one of Dodd's addresses dealing with fossil monotremes, or egg-laying mammals. In his inimitable way, Bram had suggested that the problem which came first, the egg or the chicken, was now seen to be linked up with the Darwinian theory, and solved in the person of Dodd.

Nevertheless, Jimmy Dodd entertained a devoted admiration for the memory of the dead scientist. He believed that Bram must have left records of inestimable importance in a cairn before he died. He wanted to find that cairn.

And he knew, what a number of Bram's enemies knew, that the dead scientist had been a morphine addict. He believed that he had wandered out into the snow under the influence of the drug.

Dodd, who shared a tent with Tommy, had raved the greater part of the night about the find.

"Well, but see here, Jimmy, suppose these beetles did inhabit the antarctic continent a few million years ago, why get excited?" Tommy had asked.

"Excited?" bellowed Dodd. "It opens one of the biggest problems that science has to face. Why haven't they survived into historic times? Why didn't they cross into Australia, like the opossum, by the land bridge then existent between that continent and South America? Beetles five feet in length, and practically invulnerable! What killed them off? Why didn't they win the supremacy over man?"

Jimmy Dodd had muttered till he went to sleep, and he had muttered worse in his dreams. Tommy was glad that Captain Storm had given them permission to return to the same spot next morning and look for further fossils, though his own interest in them was of the slightest.

The dogs were being harnessed next morning when the two men hopped into the plane. The thermometer was unusually high for the season, for in the south polar regions the short summer is usually at an end by March. Tommy was sweating in his furs in a temperature well above the freezing point. The snow was crusted hard, the sky overcast with clouds, and a wind was blowing hard out of the south, and increasing in velocity hourly.

"A bad day for starting," said Captain Storm. "Looks like one of the autumn storms was blowing up. If I were you, I'd watch the weather, Tommy."

Tommy glanced at Dodd, who was huddled in the rear cockpit, fuming at the delay, and grinned whimsically. "I guess I can handle her, Captain," he answered. "It's only an hour's flight to where he found that fossil."

"Just as you please," said Storm curtly. He knew that Tommy's judgment as a pilot could always be relied upon. "You'll find us here when you return," he added. "I've counter-manded the order to march. I don't like the look of the weather at all."

Tommy grinned again and pressed the starter. The engine caught and warmed up. One of the men kicked away the blocks of ice that had been placed under the skids to serve as chocks. The plane taxied over the crusted snow, and took off into the south.

The camp was situated in a hollow among the ice-mountains that rose to a height of two or three thousand feet all around. Tommy had not dreamed how strongly the gale was blowing until he was over the top of them. Then he realized that he was facing a tougher proposition than he had calculated on. The storm struck the biplane with full force.

A snowstorm was driving up rapidly, blackening the sky. The sun, which only appeared for a brief interval every day, was practically touching the horizon as it rose to make its minute arc in the sky. A star was visible through a rift in the clouds overhead, and the pale daylight in which they had started had already become twilight.

Tommy was tempted to turn back, but it was only a hundred miles, and Jimmy Dodd would give him no peace if he did so. So he put the plane's nose resolutely into the wind, watching his speed indicator drop from a hundred miles per hour to eighty, sixty, forty—less.

The storm was beating up furiously. Of a sudden the clouds broke into a deluge of whirling snow.

In a moment the windshield was a frozen, opaque mass. Tommy opened it, and peered out into the biting air. He could see nothing.... The plane, caught in the fearful cross-currents that swirl about the southern roof of the world, was fluttering like a leaf in the wind. The altimeter was dropping dangerously.

Tommy opened the throttle to the limit, zooming, and, like a spurred horse, the biplane shot forward and upward. She touched five thousand, six, seven—and that, for her, was ceiling under those conditions, for a sudden tremendous shock of wind, coming in a fierce cross-current, swung her round, tossed her to and fro in the enveloping white cloud. And Tommy knew that he had the fight of his life upon his hands.

The compasses, which required considerable daily adjusting to be of use so near to the pole, had now gone out of use altogether. The air speed indicator had apparently gone west, for it was oscillating between zero and twenty. The turn and bank indicator was performing a kind of tango round the dial. Even the eight-day clock had ceased to function, but that might have been due to the fact that Tommy had neglected to wind it. And the oil pressure gauge presented a still more startling sight, for a glance showed that either there was a leak or else the oil had frozen.

Tommy looked around at Dodd and pointed downward. Dodd responded with a vicious forward wave of his hand.

Tommy shook his head, and Dodd started forward along the cabin, apparently with the intention of committing assault and battery upon him. Instead, the archaeologist collapsed upon the floor as the plane spun completely around under the impact of a blast that was like a giant's slap.

The plane was no longer controllable. True, she responded in some sort to the controls, but all Tommy was able to do was to keep her from going into a crazy sideslip or nose dive as he fought with the elements. And those elements were like a devil unchained. One moment he was dropping like a plummet, the next he was shooting up like a rocket as a vertical blast of air caught the plane and tossed her like a cork into the invisible heavens. Then she was revolving, as if in a maelstrom, and by degrees this rotary movement began to predominate.

Round and round went the plane, in circles that gradually narrowed, and it was all Tommy could do to swing the stick so as to keep her from skidding or sideslipping. And as he worked desperately at his task Tommy began to realize something that made him wonder if he was not dreaming.

The snow was no longer snow, but rain—mist, rather, warm mist that had already cleared the windshield and covered it with tiny drops.

And that white, opaque world into which he was looking was no longer snow but fog—the densest fog that Tommy had ever encountered.

Fog like white wool, drifting past him in fleecy flakes that looked as if they had solid substance. Warm fog that was like balm upon his frozen skin, but of a warmth that was impossible within a few miles of the frozen pole.

Then there came a momentary break in it, and Tommy looked down and uttered a cry of fear. Fear, because he knew that he must be dreaming.

Not more than a thousand feet beneath him he saw patches of snow, and patches of—green grass, the brightest and most verdant green that he had ever seen in his life.

He turned round at a touch on his shoulder. Dodd was leaning over him, one hand pointing menacingly upward and onward.

"You fool," Tommy bellowed in his ear, "d'you think the south pole lies over there? It's here! Yeah, don't you get it, Jimmy? Look down! This valley—God, Jimmy, the south pole's a hole in the ground!"

And as he spoke he remembered vaguely some crank who had once insisted that the two poles were hollow because—what was the fellow's reasoning? Tommy could not remember it.

But there was no longer any doubt but that they were dropping into a hole. Not more than a mile around, which explained why neither Scott nor

Amundsen had found it when they approximated to the site of the pole. A hole—a warm hole, up which a current of warm air was rushing, forming the white mist that now gradually thinned as the plane descended. The plateau with its covering of eternal snows loomed in a white circle high overhead. Underneath was green grass now—grass and trees!

The fog was nearly gone. The plane responded to the controls again. Tommy pushed the stick forward and came round in a tighter circle.

And then something happened that he had not in the least expected. One moment he seemed to be traveling in a complete calm, a sort of clear funnel with a ring of swirling fog outside it—the next he was dropping into a void!

There was no air resistance—there seemed hardly any air, for he felt a choking in his throat, and a tearing at his lungs as he strove to breathe. He heard a strangled cry from Dodd, and saw that he was clutching with both hands at his throat, and his face was turning purple.

The controls went limp in Tommy's hands. The plane, gyrating more slowly, suddenly nosed down, hung for a moment in that void, and then plunged toward the green earth, two hundred feet below, with appalling swiftness.

Tommy realized that a crash was inevitable. He threw his goggles up over his forehead, turned and waved to Dodd in ironic farewell. He saw the earth rush up at him—then came the shattering crash, and then oblivion!

CHAPTER II
BEETLES AND HUMANS

How long he had remained unconscious, Tommy had no means of determining. Of a sudden he found himself lying on the ground beside the shattered plane, with his eyes wide open.

He stared at it, and stared about him, without understanding where he was, or what had happened to him. His first idea was that he had crashed on the golf links near Mitchell Field, Long Island, for all about him were stretches of verdant grass and small shrubby plants. Then, when he remembered the expedition, he was convinced that he had been dreaming.

What brought him to a saner view was the discovery that he was enveloped in furs which were insufferably hot. He half raised himself and succeeded in unfastening his fur coat, and thus discovered that apparently none of his bones was broken.

But the plane must have fallen from a considerable height to have been smashed so badly. Then Tommy discovered that he was lying upon an extensive mound of sand, thrown up as by some gigantic mole, for burrow tracks ran through it in every direction. It was this that had saved his life.

Something was moving at his side. It was half-submerged in the sand-pile, and it was moving parallel to him with great rapidity.

A grayish body, half-covered with grains of sand emerged, waving two enormously long tentacles. It was a shrimp, but fully three feet in length, and Tommy had never before had any idea what an unpleasant object a shrimp is.

Tommy staggered to his feet and dropped nearer the plane, eyeing the shrimp with horror. But he was soon relieved as he discovered that it was apparently harmless. It slithered away and once more buried itself in the pile of sand.

Now Tommy was beginning to remember. He looked into the wreckage of the plane. Jim Dodd was not there. He called his name repeatedly, and there was no response, except a dull echo from the ice-mountains behind the veil of fog.

He went to the other side of the plane, he scanned the ground all about him. Jimmy had disappeared. It was evident that he was nowhere near, for Tommy could see the whole of the lower scope of the bowl on every side of him. He had walked away—or he had been carried away! Tommy thought of the shrimp, and shuddered. What other fearsome monsters might inhabit that extraordinary valley?

He sat down, leaning against the wreck of the fuselage, and tried to adjust his mind, tried to keep himself from going mad. He knew now that the flight had been no dream, that he was a member of his uncle's expedition, that he had flown with Jim toward the pole, had crashed in a vacuum. But where was Jim? And how were they going to get out of the damn place?

Something like a heap of stones not far away attracted Tommy's attention. Perhaps Jim Dodd was lying behind that. Once more Tommy got upon his feet and began walking toward it. On the way, he stumbled against the sharp edge of something that protruded from the ground.

It cut his leg sharply, and, with a curse, he began rubbing his shin and looking at the thing. Then he saw that it was another of the fossil shells, half-buried in the marshy ooze on which he was treading. The ground in this lower part of the valley was a swamp, on account of the very fine mist falling from the fog clouds that surrounded it impenetrably on every side.

Then Tommy came upon another shell, and then another. And now he saw that there were piles of what he had taken to be rock everywhere, and that this was not rock but great heaps of the shells, all equally intact.

Hundreds of thousands of the prehistoric beetles must have died in that valley, perhaps overcome by some cataclysm.

Tommy examined the heap near which he stood; he yelled Dodd's name, but again no answer came.

Instead, something began to stir among the heaps of shells. For a moment Tommy hoped against hope that it was Dodd, but it wasn't Dodd.

It was a living beetle!

A beetle fully five feet high as it stood erect, a pair of enormous wings outspread. And the head, which was larger than a man's, was the most frightful object Tommy had ever seen.

Jim Dodd would have said at once that this was one of the Curculionidae, or snout beetles, for a prolongation of the head between the eyes formed a sort of beak a foot in length. The mouth, which opened downward, was armed with terrific mandibles, while the huge, compound eyes looked like enormous crystals of cut glass. Immediately in front of the eyes were two mandibles as long as a man's arms, with feathery processes at the ends. In addition to these there were three pairs of legs, the front pair as long as a man's, the hind pair almost as long as a horse's.

Paralyzed with horror, Tommy watched the monster, which had apparently been disturbed by the vibrations of his voice, extract itself from among the shells. Then, with a bound that covered fifteen feet, it had lessened the distance between them by half.

And then a still more amazing thing happened. For of a sudden the hard shell slipped from the thorax, the wing-cases dropped off, the whole of the bony parts slipped to the ground with a clang, and a soft, defenseless thing went slithering away among the rocks.

The beetle had moulted!

Tommy dropped to the ground in the throes of violent nausea.

Then, looking up again, he saw the girl!

She was about a hundred yards away from him, very close to the fallen plane, and she must have emerged from a large hole in the ground which Tommy could now see under a ledge of overhanging rock.

She seemed to be dressed in a single garment which fell to her knees, and appeared to fit tightly about her body, but as she came nearer, Tommy, watching her, petrified by this latest apparition, discovered that it was woven of her own hair, which must have been of immense length, for it fell naturally to her shoulders, and thence was woven into this close-fitting material, a fringe an inch or two in length extending beneath the selvage.

She was about six feet tall, and apparently made after the normal human pattern. She moved with a slow, majestic swing, and if ever any female had seemed to Tommy to have the appearance of an angel, this unknown woman did.

She was so fair, in that flossy, flaxen covering, she moved with such easy grace, that Tommy, gaping, gradually crept nearer to her. She did not seem to see him. She was stooping over the very sand heap into which he had fallen. Suddenly, with lightning-like rapidity, her arms shot out, her hands began tunneling in the sand. With a cry of triumph she pulled out the shrimp Tommy had seen, or another like it, and, stripping it off the shell, began devouring it with evident relish.

In the midst of her meal the girl raised her head and looked at Tommy. He saw that her eyes were filmed, vacant, dead. Then of a sudden a third membrane was drawn back across the pupils, and she saw him.

She let the shrimp drop to the ground, uttered a cry, and moved toward him with a tottering gait. She groped toward him with outstretched arms. And then she was blind again, for the membrane once more covered her pupils. It was as if her eyes were unable to endure even the dim light of the valley, through whose surrounding mists the low sun, setting just above the horizon, was unable to diffuse itself save as a brightening of the fog curtain.

Tommy stepped toward the girl. His outstretched hand touched hers. It was unquestionably a woman's hand he held, delicately warm, with exquisitely moulded fingers, in whose touch there seemed to be, for the girl, some tactile impression of him.

Again that membrane was drawn back from the girl's pupils for a fleeting flash. Tommy saw two eyes of intense black, their color contrasting curiously with the flaxen color of her hair and her white skin, almost the tint of an albino's. Those eyes had surveyed him, and appeared satisfied that he was one of her kind. She could not have seen very much in that almost instantaneous flash of vision. Queer, that membrane—as if she had been used to living in the dark, as if the full light of the day was unbearable!

She drew her hand away. Soft vocals came from her lips. Suddenly she turned swiftly. She could not have seen, but before Tommy had seen, she had sensed the presence of the old man who was creeping out of the hole in the mountainside.

He moved forward craftily, and then pounced upon the sand pile, and in a moment had pulled out another of the big shrimps, which he proceeded to devour with greedy relish. The girl, leaving Tommy's side, joined him in that unpleasant feast.

And in the midst of it a flood came pouring from the hole—a flood of living beetles, covering the ground in fifteen-foot leaps as they dashed at the two.

To his horror, Tommy saw Jimmy Dodd among them, wrapped in his fur coat like a mummy, and being pushed and rolled forward like a football.

For a moment Tommy hesitated, torn between his solicitude for Jim Dodd and that for the girl. Then, as the foremost of the monsters bounded to her side, he ran between them. The vicious jaws snapped within six inches of Tommy's face, with a force that would have carried away an ear, or shredded the cheek, if they had met.

Tommy struck out with all his might, and his fist clanged on the resounding shell so that the blood spurted from his bruised knuckles. He had struck the monster squarely upon the thorax, and he had not discommoded it in the least. It turned on him, its glassy, many-faceted eyes glaring with a cold, infernal light. Tommy struck out again with his left hand, this time upon the pulpy flesh of the downward-opening mouth.

An inch higher, and he would have impaled his hand upon the beak, with a point like a needle, and evidently used for purposes of attack, since it was not connected with the mandibles. The blow appeared to fall in the only vulnerable place. The monster dropped upon its back and lay there, unable to reverse itself, its antenna and forelegs waving in the air, and the rear legs rasping together in a shrill, strident shriek.

Instantly, as Tommy darted out of the way, the swarm fell upon the helpless monster and began devouring it, tearing strips of flesh from the lower shell, which in the space of a half-minute was reduced simply to bone. The most horrible feature of this act of cannibalism was the complete silence with which it was performed, except for the rasping of the dying monster's legs. It was evident that the huge beetles had no vocal apparatus.

For the moment left unguarded, Jim Dodd flung down the collar of his fur coat, stared about him, and recognized Tommy.

"My God, it's you!" he yelled. "Well, can you—?"

He had no time to finish his sentence. A pair of antenna went round his neck from behind. At the same instant Tommy, the old man, and the girl were gripped by the monsters, which, forming a solid phalanx about them, began hustling them in the direction of the hole. Resistance was utterly impossible. Tommy felt as if he was being pushed along by a moving wall of stone.

Inside the opening it was completely dark. Tommy shouted to Dodd, but the strident sounds of the moving legs drowned his cries. He was being pushed forward into the unknown.

Suddenly the ground seemed to fall away beneath his feet. He struggled, cried out, and felt himself descending through the air.

For a full half-minute he went downward at a speed that constricted his throat so that he could hardly draw breath. Then, just as he had nerved himself for the imminent crash, the speed of his descent was checked. In another moment he found that he was slowing to a standstill in mid-air.

He was beginning to float backward—upward. But the wall of moving shells, pushing against him, forced him on, downward, and yet apparently against the force of gravitation.

Then of a sudden Tommy was aware of a dim light all about him. His feet touched earth and grass as softly as a thistledown alighting.

He found himself seated in the same dim light upon red grass, and staring into Jimmy's face.

CHAPTER III
TEN MILES UNDERGROUND

"What I was going to say when we were interrupted, was, 'Can you beat it?'" Jimmy Dodd observed, with admirable sang-froid.

They were still seated on the red grass, gazing about them at what looked like an illimitable plain, and upward into depths of darkness. It was warm, and the light, furnished by what appeared to be luminous vegetation, was about that of twilight.

On every side were clumps of trees and shrubs, which formed centers of phosphorescent illumination, but for the most part the land was open, and here and there human figures appeared, moving with head down and arms hanging earthward.

"No, I'm damned if I can," said Tommy. "What happened to you after we crashed?"

"Why, first thing I knew, I found myself riding on the back of a fossil beetle, apparently one of the *curculionidae*," said Dodd.

"Never, mind being so precise, Jimmy. Let's call it a beetle. Go on."

"They set me down inside the hole and seemed to be investigating me, the whole swarm of them. Of course, I thought I was dead, and come to my just reward, especially when I saw those beaks. Then one of them began tickling my face with its antenna, and I drew up my fur collar. They didn't seem to like the feel of the fur, and after a while the whole gang started hustling me back again, like a nest of ants carrying something they don't want outside their hill. And then you bobbed up."

"Well, my opinion is you saved your life by pulling up your collar," said Tommy. "Looks to me as if it's a case of the survival of the fittest, said fittest being the insect, and the human race taking second place. You know what the humans here live on, don't you?"

"No, what?"

"Shrimps as big as poodles. If you'd seen that girl and the old man getting outside them, you'd realize that there seems to be a food shortage in this part of the world. Say, where in thunder are we, Jimmy?"

"Haven't you guessed yet, Travers?" asked Dodd, a spice of malice in his voice.

"I suppose this is some sort of big hole on the site of the south pole, with warm vapors coming up. Maybe a great fissure in the earth, or something."

Jimmy Dodd's grin, seen in the half-light, was rather disconcerting. "How far do you think we dropped just now?" Dodd asked.

"Why, I'd say several hundred yards," replied Tommy. "What's your estimate?"

"Just about ten miles," answered Dodd.

"What? You're still crazy! Why, we slowed up!"

"Yeah," grinned Dodd, "we slowed up. We're inside the crust of the world. That's the long and short of it. The earth we've known is just a shell over our heads."

"Yeah? Walking head downward, are we? Then why don't we drop to the center of the earth, you damn fool?"

"Because, my dear fellow, you can swing a pailful of water round your head without spilling any of it. In other words, our old friend, centrifugal force. The speed with which the earth is rotating, keeps us on our feet, head downward. To be precise, the center of the earth's gravity lies in the middle of the hollow sphere, of course, but the counteraction of centrifugal force throws it outward to the middle of the ten-mile crust. That's why we slowed down after we were half-way through. We were moving against gravity."

"And what's up there, or down there, or whatever you call it?" asked Tommy, pointing to what ought to have been the sky.

"Nothing. It's the center of the tennis ball, though I imagine it's pretty near a vacuum when you get up a mile or so, owing to the speed of the earth's rotation, which forces the heat into the shell."

"You mean to say you actually believe that stuff you've been handing me?" asked Tommy, after a pause. "Then how did human beings get here, and those damn beetles? And why's the grass red?"

"The grass is red because there's no sunlight to produce chlorophyll. The inhabitants of the deep sea are red or black, almost invariably. In the case of the humans, they've become bleached. My belief is that that man and woman we saw, and those"—he pointed to the vague forms of human beings, who moved across the grass, gathering something—"are survivors of the primitive race that still exists as the Australians. Undoubtedly one of the branches of the human stock originated in antarctica at a time when it enjoyed a tropical temperature, and was the land bridge between Australia and South America."

"And the—beetles?" asked Tommy.

"Ah, they go back to the days when nature was in a more grandiose mood!" replied the archaeologist enthusiastically. "That's the most wonderful discovery of the ages. The world will go crazy over them when we bring back the first living specimens to the zoological parks of the great cities.

"But," Dodd went on, speaking with still more enthusiasm, "of course, this is only the beginning, Tommy. There are ten million species of insects, according to Riley, and it is inevitable that there must be hundreds of thousands of other survivals from the age of the great saurians, perhaps even some of the saurians themselves. Who knows but that we may discover the ancestor of the extinct monotremes, the rhynchocephalia, the pterodactyls, hatch a brood of aepyornis eggs—"

"And," said Tommy tartly, "how are we going to get them back, apart from the little problem of getting out of here ourselves?"

"Don't let's worry about that now," answered Dodd. "It will take ten years of the hardest kind of labor even to begin a classification of the inhabitants of this inner world. I could sit down for ever, and—"

But Jimmy Dodd rose to his feet as a pair of antenna whipped round his neck and jerked him bodily upward.

One of the monster beetles was standing upright behind them, and by its gestures it evidently meant that Dodd and Tommy were to join the crowd of humans in the offing. As Dodd turned upon it with an indignant show of fists, one of the antennae whipped off his fur coat and stung him painfully with the bristle-like attachment at the end.

It was a painful moment when Dodd and Tommy realized that they were powerless against the monstrous beetles. Tommy tried the uppercut with which he had knocked out the deceased monster, but the quick jerks of the present beetle's head were infinitely faster than the movements of his fists, while the antenna had a whiplike quality about them that speedily convinced him that discretion was the card to play.

Under the threat of the curling antenna, Tommy and Dodd moved in the direction of the slowly circulating humans. Numerous tiny rodents, which evidently kept the red grass short, scampered away under their feet. The beetles made no further effort to force them on, but now they could see that a number of the monsters were stationed at intervals around a wide circle, keeping the humans in a single body.

"Good Lord!" ejaculated Tommy, stopping. "See what they're doing, Dodd? They're herding us, like cowboys herd steers. Look at that!"

One of the herd, a male with a long beard, suddenly broke from the herd, bawling, and flung himself upon a beetle guard. The antenna shot forth, coiled around his neck, and hurled him a dozen feet to the ground, where he lay stunned for a moment before arising and rejoining his companions.

"But what are they looking for?" demanded Dodd.

Tommy had not heard him. He had stopped in front of one of the luminous trees and was plucking a fruit from it.

"Jimmy, ever see an apple before?" he asked. "If this isn't an apple, I'll eat my head."

It certainly was an apple, and one of the largest and juiciest that Tommy had ever tasted. It was the reddest apple he had ever seen, and would have won the first prize at any agricultural fair.

"And look at this!" shouted Tommy, plucking an enormous luminous peach from another tree.

They began munching slowly, then, seeing one of the beetle guards approaching them, they moved into the midst of the crowd.

"Did you notice anything strange about those fruit trees?" inquired Dodd, as he munched. "I'll swear they were monocotyledonous, which, after all, is what one would expect. Still, to think that the monocotyledons evolved the familiar drupes, or stone fruits, on a parallel line to the dicotyledons is—amazing!"

A box on the ear like the kick of a mule's hoof jerked the last word from his lips as he went sprawling. He got up, to see the girl standing before him, intense disgust and anger on her face.

She snatched the fruits from the hands of the two Americans and hurled them away. It was evident from her manner that she considered such diet in the highest degree unclean and disgusting; also that she considered herself charged with the duty of superintending Tommy's and Dodd's education, but especially Dodd's.

Taking him by the arm, she propelled him into the midst of the groping humans. She released him, stooped, and suddenly stood up, a shrimp about eighteen inches long in her hand.

Towering over Dodd by six inches, she took his face in her hands and began caressing him; then, seizing his jaws in her strong fingers, she pried them apart, and popped the tail end of the shrimp into his mouth.

Dodd let out a yelp, and spat out the love-gift, to be rewarded with another box on the ear by the young Amazon, while Tommy stood by, convulsed with laughter, and yet in considerable trepidation, for fear of being forced to share Dodd's fate.

For the girl was again holding out the tail end of the crustacean, and Jim Dodd's jaws were slowly and reluctantly approaching it.

But suddenly there came an intervention as the strident rasping of beetle legs was heard in the distance. Panic seized the human herd, grovelling for shrimps in the sandy soil with its tufts of red grasses. Milling in an uneasy mob, they cowered under the lashes of the antenna of the beetle guards, which sacrificed their backs through their hair garments whenever any of them tried to bolt.

Nearer and nearer came the beetles, louder and more penetrating the shriek of their rasping legs. Now the swarm came into sight, rank after rank of the shell-clad monsters, leaping fifteen feet at a bound with perfect precision, until they had formed a solid phalanx all around the humans.

Tommy heard sighs of despair, he heard muttering, and then he realized, with deep thankfulness, that these human beings, degraded though they were, had a speech of their own.

In the middle of the front line appeared a beetle a foot taller than the rest. That it was either a king or queen was evident from the respect paid it by the rest of the swarm. At its every movement a bodyguard of beetles moved in unison, forming themselves in a group before it and on either side.

There would have been something ludicrous about these movements, but for the impression of horror that the swarm made upon Tommy and Jim Dodd. Hitherto both had supposed that the hideous insects acted by blind instinct, but now there could no longer be any doubt that they were possessed of an organized intelligence.

The strident sounds grew louder. Already Tommy was beginning to discover certain variations in them. It was dawning upon him that they formed a language—and a perfectly intelligible one. For, as the note changed about a half-semitone, two of the monsters left the side of their ruler and reached the two men with three successive leaps.

Their movements left no doubt in either Tommy's or Dodd's mind what was required. The two strode hastily toward the assemblage, and stopped as the antenna of their guards came down in menacing fashion.

It was light enough for Tommy to see the face of the ruler of the hellish swarm. And it required all his powers of will to keep from collapsing from sheer horror at what he saw.

For, despite the close-fitting shell, the face of the beetle king was the face of a man—a white man!

Jim Dodd's shriek rang out above the shrilling of the beetle-legs, "Bram! It's you, it's you! My God, it's you, Bram!"

CHAPTER IV
Bram's Story

A sneering chuckle broke from Bram's lips. "Yes, it's me, James Dodd," he answered. "I'm a little surprised to see you here, Dodd, but I'm mighty glad. Still insane upon the subject of fossil monotremes, I suppose?"

The words came haltingly from Bram's lips, as from those of a man who had lost the habit of easy speech. And Tommy, looking on, and trying to keep in possession of his faculties, had already come to the conclusion that the sounds were inaudible to the beetles. Probably their hearing apparatus was not attuned to such slow vibrations of the human voice.

Also he had discovered that Bram was wearing the discarded shell of one of the monsters: he had not grown the shell himself. It was fastened about his body by a band of the hair-cloth, fastened to the two protuberances of the elytra, or wing-cases, on either side of the dorsal surface.

The discovery at least robbed the situation of one aspect of terror. Bram, however he had obtained control of the swarm, was still only a man.

"Yes, still insane," answered Dodd bitterly. "Insane enough to go on believing that the polyprotodontia and the dasyuridae, which includes the peramelidae, or bandicoots, and the banded ant-eaters, or myrmecobidae, are not to be found in fossil form, for the excellent reason that they were not represented before the Upper Cretaceous period."

"You lie! You lie!" screamed Bram. "I have shown to all the world that phascalotherium, amphitherium, amblotherium, spalacotherium, and many other orders are to be found in the Upper Jurassic rocks of England, Wyoming, and other places. You—you are the man who denied the existence of the nototherium, of the marsupial lion, in pleistocene deposits! You denied that the dasyuridae can be traced back beyond the pleistocene. And you stand there and lie to me, when you are at my mercy!"

"For God's sake don't aggravate him," whispered Tommy to Dodd. "Don't you see that he's insane? Humor him, or we'll be dead men. Think what the world will lose, if you are never able to go back with your specimens," he added craftily.

But Dodd, whose eyes were glaring, said a sublime thing: "I have given my life to science, and I will never deny my master!"

With a screech, which, however, was evidently inaudible to the beetles, Bram leaped at Dodd and seized him by the throat. The two men fell to the ground, the ponderous beetle-shell completely covering them. Underneath it they could be seen to be struggling desperately. All the while the beetle horde remained perfectly motionless. Tommy thought afterward that in this fact lay their brightest chances of escape, if Bram's immediate vengeance did not fall on them.

Either because Bram was not himself a beetle, or because in some other way the swarm instinct was not stirred, the monsters watched the struggle with complete indifference.

At the moment, however, Tommy was only concerned with saving Dodd from the madman. He got his foot beneath the shell, then inserted his leg; using his whole body as a lever, he succeeded in turning Bram over on his back.

Then, and only then, the swarm rushed in upon them. Then Tommy realized that he had touched one of the triggers that regulated the beetle's automatism. In another instant Bram would have been torn to pieces. The needle-beaks were darting through the air, the hideous jaws were snapping. Bram's yells rang through the cavern.

Dodging beneath the avalanche of the monsters, Tommy got Bram upon his feet again. The beetles stopped, every movement arrested. Bram's hand went to the pocket of his tattered coat, there came a snap, a flash. Bram had ignited an automatic cigarette-lighter!

Instantly the monsters went scurrying away into the distance. And Tommy had another clue. The beetles, living in the dimness of the underworld, could not stand light or fire!

He ran to where Jimmy was lying, face upward, on the ground. His face was badly scarred by Bram's nails, and the blood was spurting from a long gash in his throat, made by the sharp flint that was lying beside him.

He had some time before discarded his fur coat. Now he pulled off his coat, and, tearing off the tail of his shirt, he made a pad and a bandage, with which he attempted to staunch the blood and bind the wound. It must have taken ten minutes before the failing heart force enabled him to get the bleeding under control. Dodd had nearly bled to death, his face was drawn and waxen, but, because the pulsation was so feeble, the artery had ceased to spurt.

Then only did Tommy take notice of Bram. He had been squatting near, and Tommy realized that he had unconsciously observed Bram put some sort of pellets into his mouth. Now he realized that Bram was a drug fiend. That was what had made him walk out of the Greystoke camp in the storm.

Bram got up and came toward them. "Is he dead?" he whispered hoarsely. "I—I lost my temper. You two—I don't intend to kill you. There—there's room for the three of us. I've got—plans of the utmost importance to humanity."

"I don't think much of the way you've started to carry them out," answered Tommy bitterly. "No, he's not dead yet, but I wouldn't give much for his chances, even in the best hospital. The best thing you can do now is to go to hell, and take your beetles with you," he added.

Bram, without replying, raised his head and emitted from his throat the shrillest whistle that Tommy had ever heard. The response was amazing.

Rasping out of the darkness came eight beetles in pairs. Instead of leaping from an upright position, they trotted in the manner of horses, on all fours, their shells, which touched at the edges, forming a solid surface, gently rounded in the center so that a man's body could lie there and fit snugly into the groove.

"Help me get him up," said Bram. "Trust me! I'll do my best for him. If we leave him here they may kill and eat him. I can't trust all those beetle guards."

Tommy hesitated a moment, then decided to follow Bram's suggestion. Together they raised the unconscious man to the beetle-shell couch. Bram seated himself upon the boss of one of the beetle-shells in front, and Tommy jumped up behind.

Next moment, to his amazement, the trained steeds were flying smoothly through the air, at a rate that could not have been less than seventy-five to eighty miles an hour.

Tommy's shell seat was not a bed of roses, but he hardly noticed that. He was thinking that if Dodd lived they should be able to turn the tables.

For, unknown to Bram, he was in possession of the cigarette-lighter which he had picked up, and which Bram, in his agitation, had forgotten. It was full of petrol, or some other fluid of a similar nature, which Bram must have obtained from some natural source within the earth. And, in an emergency, Tommy knew that he had the means of keeping the beetles at bay.

They had traveled for perhaps an hour when a faint light began to glow in the distance. It grew brighter, and a roaring sound became audible. A turn of the track that they were traversing, and the light became a glare. A terrific sight met Tommy's eyes.

Out of the bowels of the earth—actually out of the crust beneath their feet—there shot a pillar of roaring flame, of intense white color, and radiating a heat that was perceptible even at a distance of several hundred yards. The beetle steeds dropped gently to the ground; they halted. Bram got down, grinning.

"Nicely trained horses, what?" he asked. "By the way, you have the advantage of me in names. Who and what are you?"

Tommy told him.

"Well, Travers, it looks as if we're going to be companions for some time to come, and I quite admit you saved my life back there. So we don't want to start with secrets. This is a natural petrol spring, which has probably been burning undiminished for ages. My trained beetles are blind—you didn't happen to notice I'd cut off their antenna? But the rest of the swarm daren't come near it. So that makes me their master.

"Pretty trick, what, Travers? I'm the Lord of the Flame down here, and I'm using my advantage. But don't get the idea of supplanting me. There are lots of other tricks you don't know anything about, and I'll have to trust you better before—"

He broke off and slipped another pellet into his mouth.

"Help me get Dodd down, if this is our destination," answered Tommy.

They lifted Dodd to the ground. He was conscious now, and moaning for water. The two men carried him into a sort of large cavern, at the farther end of which the fire was roaring. Bram went to a spring that trickled down one side, filled something that looked like a petrified lily calyx, and brought it to Dodd, who drained it.

Tommy looked about him. He was astonished to see that the place was, in a way, furnished. Bram had carved out a very creditable couch, and several low chairs, evidently with a stone ax, for by the light of the fire, which cast a fair illumination even at that distance, Tommy could see the marks of the implement, rough and irregular, in the wood.

On the ground were thick rugs, woven of hair, and two or three more rugs of the same material lay on the couch. It was evident that the human herd was expected to furnish textile materials as well as meat.

"Sit down, and make yourself comfortable," said Bram, when they had raised Dodd to the couch. "We'll have dinner, and then we'll talk. I can give you a fine vegetarian meal. Those dirty shrimp-eating savages look on me as a cannibal because I eat the fruits of the trees." He grinned. "There's a bad shortage of food in Submundia, as I've named this part of the world," he went on, "for until I came the beetles simply devoured the humans wholesale, instead of breeding them, like I taught them. And there's another of the hundred-and-fifty year swarms due to hatch out soon. However, we'll talk about that later. And all those fine fruits going to waste! Excuse me, Travers."

He disappeared, and returned in a minute or two with a small table, piled high with luscious fruits unknown to Tommy, though among them were some that looked like loaves of natural bread.

Tommy, whose appetite never failed him even in the worst circumstances, fell to with a will. He was enjoying his meal when he happened to look up, and saw that the penumbra at the edge of the lighted zone was dense with beetles.

Thousands—perhaps millions, for they stretched away as far as the eye could see, were packed together, their antenna waving in unison, their heads, beneath the shells, directed toward the fire.

Bram saw Tommy's look of disgust, and laughed. "The fire seems to intoxicate them, Travers," he said. "They always throng the entrance when I'm here. It's as far as they dare go. They're quite blind in the least light. Care to smoke? I've learned the art of making some quite decent cigars." He produced a handful. "Oh, by the way, you didn't see my lighter anywhere, did you?" he went on, with a pretense of carelessness.

"No," lied Tommy. "I was surprised you—"

"Oh, there's a supply of petrol in the rocks. No matter," answered Bram carelessly. "Your friend looks bad," he added, glancing at Dodd, who had fallen asleep. "Travers, I'm sorry I lost my temper. The—the shock of meeting men from the upper world, you know."

Dodd opened his eyes and tried to whisper. Tommy bent over him and listened.

"He wants to know whether he can have that girl to take care of him," he said.

"What, the one I saw you with? Why, she's a cull, Travers."

"What d'you mean?" asked Tommy.

"Why—useless, you know. There's several of them running loose, and waiting to be rounded up. We raise two breeds, one for replenishing the stock, and one for meat. She's just a cull, a reversion, no use for either purpose. I'll have her brought by all means. I—I like Dodd. I want to get him to like me," Bram went on, with a sort of penitence that had a pathetic touch. "Our little differences—quite absurd, and I can prove he's wrong in his ideas.

"Make yourself comfortable as long as you're here, Travers, and don't mind me. Only, don't try to escape. The beetles will get you if you do, and there's no way out of here—none that you'll find. And don't try to follow me. But you're a sensible man, and we'll all get along famously, I'm sure, as soon as Dodd recovers."

CHAPTER V
DOOMED!

There were no means known to Tommy of reckoning time in that strange place of twilight. His watch had been broken in the airplane fall; and Dodd never remembered to wind his, but they estimated that about two weeks had passed, judging from the number of times they had slept and eaten.

In those two weeks they had gradually begun to grow accustomed to their surroundings. Haidia, the girl, had arrived on beetle-back within an hour after Bram's departure, apparently into a cleft of the rocks—how he had communicated his order to the beetle steeds Tommy had no idea. And under the girl's ministrations Dodd was making good progress toward recovery.

That Haidia was in love with Dodd in quite a human way was evident. To please the girl, both Dodd and Tommy had learned to eat the raw shrimps, which, being bloodless, were really no worse than oysters, and had a flavor half-way between shrimp and crawfish. To please the men, Haidia tried not to shudder when she saw them devouring the breadfruit and nectarines of which Bram always had a plentiful supply. Bram was solicitous in his inquiries for Dodd's health.

"Jim, I've been thinking about our chances of getting away," said Tommy one morning. "It's evident Bram's only waiting for your recovery to put some proposition up to us. Suppose you were to feign paralysis."

"How d'you mean? What for?" demanded Dodd.

"If he thinks you're helpless, he'll be less on his guard. You haven't walked about in his presence." That was true, for the activities of the two had been nocturnal, when Bram had vanished. "Let him think a nerve's been severed in your neck, or something of the sort. If it doesn't work, you can always get better."

Dodd's realistic portrayal of a man with a partly paralyzed right side brought cries of horror from Bram next morning. Solicitously he helped Dodd back to the couch. Bram, when not under the influence of his drug, had moments of human feeling.

"Can't you move that arm and leg at all, Dodd?" he asked. "No feeling in them?"

"There's plenty of feeling," growled Dodd, "but they don't seem to work, that's all."

"You'll get better," said Bram eagerly. "You must get better. I need you, Dodd, in spite of our differences. There's work for all of us, wonderful work. A new humanity, waiting to be born, Dodd, not of the miserable ape race, but of—of—"

He checked himself, and a cunning look came over his face. He turned away abruptly.

At the end of two weeks or so, an amazing thing happened. One day Haidia, with a look of triumph in her eyes, addressed Dodd with a few English words!

Her brain, which had probably developed certain faculties in different proportions from those of the upper human race, had registered every word that either of the two men had ever spoken, and remembered it. As soon as Dodd ascertained this, he began to instruct her, and, with her abnormal faculties of memory, it was not long before she could talk quite intelligently. The obstacle that had stood between them was swept away. She became one of themselves.

In the days that followed the girl told them brokenly something of the history of her race, of the legend of the universal flood that had driven them down into the bowels of the earth, of the centuries-long struggle with the beetles, and of the insects' gradual conquest of humanity, and the final reduction of the human race to a miserable, helpless remnant.

Everywhere, Haidia told them, were beetle swarms, everywhere humanity had been reduced to a few handfuls. Bram, by breeding mankind from prolific strains, and using the new-born progeny for food, had temporarily averted universal starvation. But a new swarm of beetles was due to hatch out shortly, and then—

The girl, with a shudder, put her hand to her bosom, and brought out a little bright-eyed lizard.

"The old man you saw with me, who is one of our wise elders, has told our people that these things feed upon the beetle larvae," she said. "We are putting them secretly into the nests. But what can a few lizards do against millions." She looked up. "In the earth above us, the beetle larvae extend for miles, in a solid mass," she said. "When they come out as beetles, it will be the end of all of us."

Bram had grown less suspicious as the time passed. His sudden visits to the cavern had ceased. Dodd and Tommy knew that he spent the nights— if they could be termed nights—lying in a drugged slumber somewhere among the rocks. They had asked Haidia whether there was any way of escape into the upper world.

"There are two ways from here," answered the girl. "One is the way you came, but it is impossible to pass the beetle guards without being torn to pieces. The other—"

She shuddered, and for an instant drew back the film from across her pupils, then uttered a little cry of pain at the light, dim though it was.

"There is a bridge across that terrible monster that devours all it touches," she said, shuddering, meaning the fire.

Suddenly Dodd had an inspiration. He still had the fur coat that he had worn, and, reaching into a pocket he drew out a pair of snow goggles, which he adjusted over Haidia's nose.

"Now look!" he said.

Haidia looked, blinked and, with an effort kept her eyes open. She gazed at Dodd in amazement. Dodd laughed, and pulled her toward him. He kissed her, and Haidia's eyes closed.

"What is this?" she murmured. "First you give me medicine that opens my eyes, and then you give me medicine that closes them."

"That's nothing," grinned Dodd. "Wait till you understand me better."

Bram's eyes were preternaturally bright. It was evident that he had been increasing his dose of late, and that he was fully under the influence of it now.

"Well, gentlemen, the time has come for us to be frank with one another," he said, as the three were gathered about the little table, while Haidia crouched in a far corner of the cave. "I want you to work for me in my plans for the regeneration of humanity. The time for which I have long labored is almost at hand. Any day now the new swarm of beetles may emerge from the pupal stage. But before I speak further, come and see them, gentlemen!"

He rose, and Dodd and Tommy rose too, Tommy supporting Dodd, who let his arm and leg trail awkwardly as he moved.

Bram led the way into the cleft among the rocks into which he had been in the habit of passing. Beyond this opening the two men saw another smaller cavern, with a beetle guard standing on either side, antenna waving.

Bram shrilled a sound, and the antenna dropped. The three passed through. Tommy saw a hair-cloth pallet set against the rocks, a table, and a chair. Beyond was a sloping ramp of earth. Overhead was a rock ceiling.

Bram led the way up the ramp, and the three stepped through a gap in the rocks and found themselves on an extensive prairie. But in place of the red grass there was a vast sea of mud.

By the light cast by the petrol fire, which roared up in the distance, a veritable fiery fountain, the two Americans could see that the mud was filled with huge encysted forms, grubs three or four feet long, motionless in the soil.

Bram scooped up one of them and tossed it into the air. It thudded to their feet and remained motionless.

"As far as you can see, and for miles beyond, these pupae of the beetles lie buried in the decaying vegetation in which the eggs were hatched," said Bram. "Every century and a half, so far as I have been able to judge from comparative anatomy, a fresh swarm emerges. See!"

He pointed to the pupa he had unearthed, which, as if stirred into activity by his handling, was now beginning to move. Or, rather, something was moving inside the cocoon.

The shell broke, and the hideous head and folded antenna of a beetle appeared. With a convulsive writhing, the monster threw off the covering and stepped out. It extended its wings, glistening, with moisture, from the still soft and pliant carapace, or shell, and suddenly zoomed off into the distance.

Tommy shuddered as the boom of its flight grew softer and subsided.

"Any day now the entire swarm will emerge," cried Bram. "How many moultings they undergo before they undergo the finished state, I do not know, but already, as you see, they are prepared for the battle of life. They emerge ravenous. That beetle will fall upon the man-herds and devour a full grown man, unless the guards destroy it."

He raised his arms with the gesture of an ancient prophet. "Woe to the human race," he cried, "the wretched ape spawn that has cast out its teachers and persecuted those who sought to raise it to higher things!"

Tommy knew that Bram was referring to himself. Bram turned fiercely upon Dodd.

"When I joined the Greystoke expedition," he cried, "it was with the express intention of refuting your miserable theories as to the fossil monotremes. I could not sleep or eat, so deeply was I affronted by them. For, if they were true, the dasyuridae are an innovation in the great scheme of nature, and man, instead of being a mere afterthought, a jest of the Creative Force, came to earth with a purpose.

"That I deny," he yelled. "Man is a joke. Nature made him when she was tired, as the architect of a cathedral fashions a gargoyle in a sportive moment. It is the insect, not man, who is the predestined lord of the ages!"

And for once in his life, perhaps because at this point Tommy dug him violently in the ribs, Dodd had the sense to remain silent. Bram led the way swiftly back into the larger cave.

"When this swarm hatches out," he said, "I calculate that there will be a trillion beetles seeking food. There is no food for a tithe of them here underneath the earth. What then? Do you realize their stupendous power, their invincibility?

"No, you don't realize it, because your minds, through long habit, are only attuned to think in terms of man. All man's long history of slaughter of the so-called lower creatures obsesses you, blinds your understanding. A beetle? Something to be trodden underfoot, crushed in sport! But I tell you, gentlemen, that nature—God, if you will—has designed to supplant the man-ape by the beetle.

"He has resolved to throw down the wretched so-called intelligence of your kind and mine, and supplant it by the divine instinct of the beetle, an instinct that is infinitely superior, because it arrives at results instantaneously. It knows where man infers. Attuned closely to nature, it alone is able to fulfil the divine plan of Creation."

Bram was certainly under the influence of his drug; nevertheless, so violent were his gestures, so inspired was his utterance, that Tommy and Dodd listened almost in awe.

"They are invincible," Bram went on. "Their fecundity is such that when the new swarm is hatched out their numbers alone will make them irresistible. They do not know fear. They shrink from nothing. And they will follow me, their leader—I, who know the means of controlling them. How, then, can puny man hope to stand against them?

"Join me, gentlemen," Bram went on. "And beware how you decide rashly. For this is the supreme moment, not only of your own lives, but for all humanity and beetledom. Upon your decision hangs the future of the world.

"For, irresistible as the beetles are, there is one thing they lack. That is the sense of historic continuity. If they destroy man, they will know nothing of man's achievements, poor though these are. My own work on the fossil monotremes—"

"Which is a tissue of inaccuracies and half-baked deductions!" shouted Dodd.

Bram started as if a whip had lashed him. "Liar!" he bawled. "Do you think that I, who left the Greystoke expedition in a howling blizzard because I knew that here, in the inner earth, I could refute your miserable impostures—do you think that I am in the mood to listen to your wretched farrago of impossibilities?"

"Listen to me," bawled Dodd, advancing with waving arms. "Once and for all, let me tell you that your deductions are all based upon fallacious premises. No, I will not shut up, Tom Travers! You want me to aid your damned beetles in the destruction of humanity! I tell you that your phascalotherium, amphitherium, and all the rest of them, including the marsupial lion, are degenerate developments of the age following the pleistocene. I say the whole insect world was made to fertilize the plant world, so that it should bear fruit for human food. Man is the summit of the scale of evolution, and I will never join in any infamous scheme for his destruction."

Bram glared at Dodd like a madman. Three times he opened his mouth to speak, but only inarticulate sounds came from his throat. And when at last he did speak, he said something that neither Dodd nor Tommy had anticipated.

"It looks as if you're not so paralysed as you made out," he sneered. "You'll change your mind within what used to be called a day, Dodd. You'll crawl to my feet and beg for pardon. And you'll recant your lying theories about the fossil monotremes, or you die—the pair of you—you die!"

CHAPTER VI
ESCAPE!

"I heard what he said. You shall not die. We shall go away to your place, where there are no beetles to eat us, even if"—Haidia shuddered—"even if we have to cross the bridge of fire, beyond which, they tell me, lies freedom."

High over and a little to one side of the petrol flame Dodd and Tommy had seen the slender arch of rock leading into another cleft in the rocks. They had investigated it several times, but always the fierce heat had driven them back.

Both Dodd and Tommy had noticed, however, that at times the fire seemed to shrink in volume and intensity. Observation had shown them that these times were periodical, recurring about every twelve hours.

"I think I've got the clue, Tommy," said Dodd, as the three watched the fiery fountain and speculated on the possibility of escape. "That flow of petrol is controlled, like the tides on earth, by the pull of the moon. Just now it is at its height. I've noticed that it loses pretty nearly half its volume at its alternating phase. If I'm right, we'll make the attempt in about twelve hours."

"Bram's given us twenty-four," said Tommy. "But how about getting Haidia across?"

"I go where you go," said Haidia, sidling up to Dodd and looking down upon him lovingly. "I do not afraid of the fire. If it burn me up, I go to the good place."

"Where's that, Haidia?" asked Dodd.

"When we die, we go to a place where it is always dark and there are no beetles, and the ground is full of shrimps. We leave our bodies behind, like the beetles, and fly about happy for ever."

"Not a bad sort of place," said Dodd, squeezing Haidia's arm. "If you think you're ready to try to cross the bridge, we'll start as soon as the fire gets lower."

"I'll be on the job," answered Haidia, unconsciously reproducing a phrase of Tommy's.

The girl glided away, and disappeared through the thick of the beetle crowd clustered about the entrance to the cavern. Tommy and Dodd had already discovered that it was through her ability to reproduce a certain beetle sound meaning "not good to eat" that the girl could come and go. They had once tried it on their own account, and had narrowly escaped the lashing tentacles.

After that there was nothing to do but wait. Three or four hours must have passed when Bram returned from his inner cave.

"Well, Dodd, have you experienced a change of heart?" he sneered. "If you knew what's in store for you, maybe you'd come to the conclusion that you've been too cocksure about the monotremes. We're slaughtering in the morning."

"That so?" asked Dodd.

"That's so," shouted Bram. "The beetles are beginning to emerge from the pupae, and they'll need food if they're to be kept quiet. We're rounding up about threescore of the culls—your friend Haidia will be among them. We've got some caged ichneumon flies, pretty little things only a foot long, which will sting them in certain nerve centers, rendering them powerless to move. Then we shall bury them, standing up, in the vegetable mould, for the beetles to devour alive, as soon as they come out of the shells. You'll feel pretty, Dodd, standing there unable to move, with the new born beetles biting chunks out of you."

Tommy shuddered, despite his hopes of their escaping. Bram, for a scientist, had a grim and picturesque imagination.

"Dodd, there is no personal quarrel between us," Bram went on. Again that note of pathetic pleading came into his voice. "Give up your mad ideas. Admit that the banded ant-eater, at least, existed before the pleistocene epoch, and everything can be settled. When you see what my beetles are going to do to humanity, you'll be proud to join us. Only make a beginning. You remember the point I made in my paper, about spalacotherium in the Upper Jurassic rocks. It would convince anybody but a hardened fanatic."

"I read your paper, and I saw your so-called spalacotherium, reconstructed from what you called a jaw-bone," shouted Dodd. "That so-called jaw-bone was a lump of chalk, made porous by water, and the rest was in your imagination. Do your worst, Bram, I'll never crucify truth to save my life. And I'll laugh at your spalacotherium when your beetles are eating me."

Bram yelled and shrieked, he stamped up and down the cavern, shaking his fists at Dodd. At last, with a final torrent of objurgation, he disappeared.

"A pleasant customer," said Tommy. "We'll have to make that bridge, Jim, no question about it, even if it means death in the petrol fire."

"Fire's dying down fast," answered Dodd. "Haidia ought to be here soon."

"If Bram hasn't got her."

"Bram got—that girl? If Bram harms a hair of her head I'll kill him with worse tortures than he's ever dreamed of," answered Dodd, leaping up, white with rage.

"You mean you—?" Tommy began.

"Love her? Yes, I love her," shouted Dodd. "She's a girl in a million. Just the sort of helpmate I need to assist me in my work when we get back. I tell you, Tommy, I didn't know what love meant before I saw Haidia. I laughed at it as a romantic notion. 'Oh lyric love, half angel and half bird!'" he quoted, beginning to stride up and down the cavern, while Tommy watched him in amazement.

And at this moment a complete beetle entered the cave. Complete, because it had a plastron, or breast-shell, as well as a back-shell, or carapace.

A double breast-shell! A new species of beetle? An executioner beetle, sent by Bram to summon them to the torture? Tommy shuddered, but Dodd, lost in his love ecstasy, was ignorant of the creature's advent.

"'Oh lyric love—'" he shouted again, as he twirled on his heel, to run smack into the monster. The crack of Dodd's head against the beetle-shell re-echoed through the cave.

The double plastron dropped, the carapace fell down: Haidia stood revealed. The lovers, folded in each other's arms, passed momentarily into a trance.

It was Tommy who separated them. "We'll have to make a move," he said. "I think the fire's as low as it ever gets. Why did you bring the shells, Haidia?"

"To save us all from the beetles," answered the girl. "When they see us in the shells, they will not know we are human. That is what makes it so hard to have to be eaten by those beetles, when they are such dumb-bells," she added, reproducing another of Tommy's words.

"Come," she continued bravely, "let us see if we can pass the fire."

The roaring fountain made the air a veritable inferno. Overhead the rocks were red-hot. A cascade of sparks tumbled in a fiery shower from the rock roof. Dodd, holding Haidia in his arms, to protect her, staggered ahead, with Tommy in the rear. Only the beetle-shells, which acted as non-conductors of the heat, made that fiery passage possible.

There was one moment when it seemed to Tommy as if he must let go, and drop into that raging furnace underneath. He heard Dodd bawling hoarsely in front of him, he nerved himself to a last effort, beating fiercely at his blazing hair—and then the heat was past, and he had dropped unconscious upon a bed of cool earth beside a rushing river.

He was vaguely aware of being carried in Dodd's arms, but a long time seemed to have passed before he grew conscious again. He opened his eyes in utter darkness. Dodd was whispering in his ear.

"Tommy, old man, how are you feeling now?" Dodd asked.

"All—right," Tommy muttered. "How's Haidia?"

"Still unconscious, poor girl. We've got to get out of here. I heard Bram yelling in the distance. He's discovered our flight. There may be another way out of the cave, and, if so, he'll stop at nothing to get us. See if you can stand, but keep your head low. There's a low roof of rock above us."

"There's water," said Tommy, listening to the roar of a torrent that seemed to be rushing past them.

"It's a stream, and I believe these shells will float and bear our weight. We've got to try. We've got to put everything to the touch now, Tommy. I'm going to lay Haidia on one of the shells, poor girl, and start her off. Then I'll follow, and you can bring up the rear."

"I'm with you," said Tommy, getting upon his feet, and uttering an exclamation of pain as, forgetful of Dodd's injunction, he let his head strike the rock roof overhead.

In the darkness he felt the outlines of his beetle-shell lying beside the torrent. He could hear Dodd in front of him, grunting as he raised Haidia's unconscious form in his arms and deposited her in her shell. Tommy got his own shell into the stream, and held it there as the waters swirled around it.

"Ready?" he heard Dodd call.

Before he could answer, there sounded from not far away, yet strangely muffled by the rocks, Bram's bellow of fury. Bram was evidently fully drugged and beside himself. Inarticulate threats came floating through the rocky chamber.

"Bram seems to have lost his head temporarily," called Dodd, laughing. "A madman, Tommy. He insists that the marsupial lion—"

"Yes, I heard you telling him about it," answered Tommy. "You handed it to him straight. However, more about the marsupial lion later. I'm ready."

"Then let 'er go," called Dodd, and his words were swallowed up by the sound of the hollow shell striking against the rocky bank as he launched his strange craft into the water.

Tommy set one foot into the hollow of his shell, and let himself go.

Instantly the shell shot forward with fearful velocity. It was all Tommy could do to balance himself, for it seemed more unstable than a canoe. Once or twice he thought he heard Dodd shouting ahead of him, but his cries were drowned in the rush of the torrent.

Suddenly a light appeared in the distance. Tommy thought it was another of the petroleum fountains, and his heart seemed to stand still. But then he gave a gasp of relief. It was a cluster of luminous fungi, ten or twelve feet tall, emitting a glow equal to that of a dozen 40-watt electric bulbs.

By that infernal light Tommy could see that the stream curved sharply. It was about fifty feet in width, and the low rock roof had receded to some fifteen feet overhead. Instead of a tunnel, there was nothing on either side of them but a vast tract of marshy ground thinly coated with the red grass.

As Tommy looked, he saw the shell that carried the unconscious body of Haidia strike the bank beside the phosphorescent growth. He could see the girl lying in the hollow of the shell, as pale as death, her eyes closed. Dodd was close behind. As the swirl of the current caught his shell, he turned to shout a warning to Tommy.

And Tommy noticed a singular thing, of which his sense of balance had already warned him, though he had hardly given conscious thought to the matter. *The river was running up-hill!*

Of course it was, since the center of gravity was in the shell of the earth, and not in the center!

But, again, the shell of the earth was under their feet!

Then Tommy hit on the solution to the problem. If the river was running up-hill, that meant that they must be near the exterior of the earth. In other words, they had passed the center of gravity: they must be within a mile or so of the exit from Submundia!

Tommy was about to shout his discovery to Dodd when his shell grounded beside the two others, at the base of the clump of fungi.

Huge, straight, hollow stems they were, with mushroom caps, and, like all fungi, fly-blown, for Tommy could see worms nearly a foot in length crawling in and out of the porous stalks. The stench from the growth was nauseating and overpowering, utterly sickening.

"Push off and let's get out of here!" Tommy called to Dodd, who was balancing his shell against the bank, and trying to peer into Haidia's face.

At that moment he caught sight of something that made his blood turn cold!

It was an insect fully fifteen feet in height, three times that of a beetle, lurking among the fungi. He saw a hugely elongated neck, a three-cornered head with a pair of tentacles, and two pairs of legs as long as a giraffe's. But what gave the added touch of horror was that the monster, balancing itself on its hind legs, had its forelegs extended in the attitude of one holding a prayer-book!

That attitude of devotion was so terrible that Tommy uttered a wild cry of terror. At the same time another cry broke from Dodd's lips.

"God, a praying mantis!" he shouted, struggling madly to push off his shell and Haidia's.

The next moment, as if shot from a catapult, the hideous monster launched itself into the air straight toward them.

(*To be concluded in the February Number.*)

The Cave of Horror

By Captain S. P. Meek

Screaming, the guardsman was jerked through the air. An unearthly screech rang through the cavern. The unseen horror of Mammoth Cave had struck again.

Dr. Bird looked up impatiently as the door of his private laboratory in the Bureau of Standards swung open, but the frown on his face changed to a smile as he saw the form of Operative Carnes of the United States Secret Service framed in the doorway.

"Hello, Carnes," he called cheerfully. "Take a seat and make yourself at home for a few minutes. I'll be with you as soon as I finish getting this weight."

Carnes sat on the edge of a bench and watched with admiration the long nervous hands and the slim tapering fingers of the famous scientist. Dr. Bird stood well over six feet and weighed two hundred and six pounds stripped: his massive shoulders and heavy shock of unruly black hair combined to give him the appearance of a prize-fighter—until one looked at his hands. Acid stains and scars could not hide the beauty of those mobile hands, the hands of an artist and a dreamer. An artist Dr. Bird was, albeit his artistry expressed itself in the most delicate and complicated experiments in the realms of pure and applied science that the world has ever seen, rather than in the commoner forms of art.

The doctor finished his task of weighing a porcelain crucible, set it carefully into a dessicator, and turned to his friend.

"What's on your mind, Carnes?" he asked. "You look worried. Is there another counterfeit on the market?"

The operative shook his head.

"Have you been reading those stories that the papers have been carrying about Mammoth Cave?" he asked.

Dr. Bird emitted a snort of disgust.

"I read the first one of them part way through on the strength of its being an Associated Press dispatch," he replied, "but that was enough. It didn't exactly impress me with its veracity, and, from a viewpoint of literature, the thing was impossible. I have no time to pore over the lucubrations of an inspired press agent."

"So you dismissed them as mere press agent work?"

"Certainly. What else could they be? Things like that don't happen fortuitously just as the tourist season is about to open. I suppose that those yarns will bring flocks of the curious to Kentucky though: the public always responds well to sea serpent yarns."

"Mammoth Cave has been closed to visitors for the season," said Carnes quietly.

"What?" cried the doctor in surprise. "Was there really something to those wild yarns?"

"There was, and what is more to the point, there still is. At least there is enough to it that I am leaving for Kentucky this evening, and I came here for the express purpose of asking you whether you wanted to come along. Bolton suggested that I ask you: he said that the whole thing sounded to him like magic and that magic was more in your line than in ours. He made out a request for your services and I have it in my pocket now. Are you interested?"

"How does the secret service cut in on it?" asked the doctor. "It seems to me that it is a state matter. Mammoth Cave isn't a National Park."

"Apparently you haven't followed the papers. It *was* a state matter until the Governor asked for federal troops. Whenever the regulars get into trouble, the federal government is rather apt to take a hand."

"I didn't know that regulars had been sent there. Tell me about the case."

"Will you come along?"

Dr. Bird shook his head slowly.

"I really don't see how I can spare the time, Carnes," he said. "I am in the midst of some work of the utmost importance and it hasn't reached the stage where I can turn it over to an assistant."

"Then I won't bother you with the details," replied Carnes as he rose.

"Sit down, confound you!" cried the doctor. "You know better than to try to pull that on me. Tell me your case, and then I'll tell you whether I'll go or not. I can't spare the time, but, on the other hand, if it sounds interesting enough...."

Carnes laughed.

"All right, Doctor," he said, "I'll take enough time to tell you about it even if you can't go. Do you know anything about it?"

"No. I read the first story half way through and then stopped. Start at the beginning and tell me the whole thing."

"Have you ever been to Mammoth Cave?"

"No."

"It, or rather they, for while it is called Mammoth Cave it is really a series of caves, are located in Edmonson County in Central Kentucky, on a spur railroad from Glasgow Junction on the Louisville and Nashville Railroad. They are natural limestone caverns with the customary stalactite and stalagmite formation, but are unusually large and very beautiful. The caves are quite extensive and they are on different levels, so that a guide is necessary if one wants to enter them and be at all sure of finding the way out. Visitors are taken over a regular route and are seldom allowed to visit portions of the cave off these routes. Large parts of the cave have never been thoroughly explored or mapped. So much for the scene.

"About a month ago a party from Philadelphia who were motoring through Kentucky, entered the cave with a regular guide. The party consisted of a man and his wife and their two children, a boy of fourteen and a girl of twelve. They went quite a distance back into the caves and then, as the mother was feeling tired, she and her husband sat down, intending to wait until the guide showed the children some sights which lay just ahead and then return to them. The guide and the children never returned."

"What happened?"

"No one knows. All that is known is the bare fact that they have not been seen since."

"A kidnapping case?"

"Apparently not, in the light of later happenings, although that was at first thought to be the explanation. The parents waited for some time. The mother says that she heard faint screams in the distance some ten minutes after the guide and the children left, but they were very far away and she isn't sure that she heard them at all. At any rate, they didn't impress her at the time.

"When half an hour had passed they began to feel anxious, and the father took a torch and started out to hunt for them. The usual thing happened; he got lost. When *he* failed to return, the mother, now thoroughly alarmed, made her way, by some uncanny sense of direction, to the entrance and gave the alarm. In half an hour a dozen search parties were on their way into the cave. The father was soon located, not far from the beaten trail, but despite three days of constant search, the children were not located. The only trace of them that was found was a bracelet which the mother identified. It was found in the cavern some distance from the beaten path and was broken, as though by violence. There were no other signs of a struggle.

"When the bracelet was found, the kidnapping theory gained vogue, for John Harrel, the missing guide, knew the cave well and natives of the vicinity scouted the idea that he might be lost. Inspired by the large reward offered by the father, fresh parties began to explore the unknown portions of the cave. And then came the second tragedy. Two of the searchers failed to return. This time there seemed to be little doubt of violence, for screams and a pistol shot were faintly heard by other searchers, together with a peculiar 'screaming howl,' as it was described by those who heard it. A search was at once made toward the spot where the bracelet had been picked up, and the gun of one of the missing men was found within fifty yards of the spot where the bracelet had been discovered. One cylinder of the revolver had been discharged."

"Were there any signs on the floor?"

"The searchers said that the floor appeared to be rather more moist and slimy than usual, but that was all. They also spoke of a very faint smell of musk, but this observation was not confirmed by others who arrived a few moments later."

"What happened next?"

"The Governor was appealed to and a company of the National Guard was sent from Louisville to Mammoth Cave. They took up camp at the mouth of the cave and prevented everyone from entering. Soldiers armed with service rifles penetrated the caverns, but found nothing. Visitors were excluded, and the guardsmen established regular patrols and sentry posts in the cave with the result that one night, when time came for a relief, the only trace that could be found of one of the guards was his rifle. It had not been fired. Double guards were then posted, and nothing happened for several days—and then another sentry disappeared. His companion came rushing out of the cave screaming. When he recovered, he admitted that both he and the missing man had gone to sleep and that he awoke to find his comrade gone. He called, and he says that the answer he received was a peculiar whistling noise which raised all the hair on the back of his neck. He flashed his electric torch all around, but could see nothing. He swears, however, that he heard a slipping, sliding noise approaching him, and he felt that some one was looking at him. He stood it as long as he could and then threw down his rifle and ran for his life."

"Had he been drinking?"

"No. It wasn't delirium either, as was shown by the fact that a patrol found his gun where he had thrown it, but no trace of the other sentry. After this second experience, the guardsmen weren't very eager to enter the cave, and the Governor asked for regulars. A company of infantry was ordered down from Fort Thomas to relieve the guardsmen, but they fared worse than their predecessors. They lost two men the first night of their guard. The regulars weren't caught napping, for the main guard heard five shots fired. They rushed a patrol to the scene and found both of the rifles which had been fired, but the men were gone.

"The officer of the day made a thorough search of the vicinity and found, some two hundred yards from the spot where the sentries had been posted, a crack in the wall through which the body of a man could be forced. This bodycrack had fresh blood on each side of it. Several of his men volunteered to enter the hole and search, but the lieutenant would not allow it. Instead, he armed himself with a couple of hand-grenades and an electric torch and entered himself. That was last Tuesday, and he has not returned."

"Was there any disturbance heard from the crack?"

"None at all. A guard was posted with two machine-guns pointed at the crack in the wall, and a guard of eight men and a sergeant stationed there. Last night, about six o'clock, while the guard were sitting around their guns, a faint smell of musk became evident. No one paid a great deal of attention to it, but suddenly for no apparent reason at all one of the men on guard was jerked into the air feet upwards. He gave a scream of fear, and an unearthly screech answered him. The guard, with the exception of one man, turned tail and ran. One man stuck by his gun and poured a stream of bullets into the crack. The retreating men could hear the rattle of the gun for a few moments and then there was a choking scream, followed by silence. When the officer of the day got back with a patrol, there was a heavy smell of musk in the air, and a good deal of blood was splashed around. The machine-guns were both there, although one of them was twisted up until it looked like it had been through an explosion.

"The Officer commanding the company investigated the place, ordered all men out of the cave, and communicated with the War Department. The Secretary of War found it too tough a nut to crack and he asked for help, so Bolton is sending me down there. Do you think, in view of this yarn, that your experiments can wait?"

The creases on Dr. Bird's high forehead had grown deeper and deeper as Carnes had told his story, but now they suddenly disappeared, and he jumped to his feet with a boyish grin.

"How soon are we leaving?" he asked.

"In two hours, Doctor. A car is waiting for us downstairs and I have reservations booked for both of us on the Southern to-night. I knew that you were coming; in fact, the request for your services had been approved before I came here to see you."

Dr. Bird rapidly divested himself of his laboratory smock and took his coat and hat from a cupboard.

"I hope you realize, Carnsey, old dear," he said as he followed the operative out of the building, "that I have a real fondness for your worthless old carcass. I am leaving the results of two weeks of patient work alone and unattended in order to keep you out of trouble, and I know that it will be ruined when I get back. I wonder whether you are worth it?"

"Bosh!" retorted Carnes. "I'm mighty glad to have you along, but you needn't rub it in by pretending that it is affection for me that is dragging you reluctantly into this mess. With an adventure like this ahead of you, leg-irons and handcuffs wouldn't keep you away from Mammoth Cave, whether I was going or not."

It was late afternoon before Dr. Bird and Carnes dismounted from the special train which had carried them from Glasgow Junction to Mammoth Cave. They introduced themselves to the major commanding the guard battalion which had been ordered down to reinforce the single company which had borne the first brunt of the affair, and then interviewed the guards who had been routed by the unseen horror which was haunting the famous cave. Nothing was learned which differed in any great degree from the tale which Carnes had related to the doctor in Washington, except that the officer of the day who had investigated the last attack failed to entirely corroborate the smell of musk which had been reported by the other observers.

"It might have been musk, but to me it smelled differently," he said. "Were you ever near a rattlesnake den in the west?"

Dr. Bird nodded.

"Then you know the peculiar reptilian odor which such a place gives off. Well, this smell was somewhat similar, although not the same by any manner of means. It was musky all right, but it was more snake than musk to me. I rather like musk, but this smell gave me the horrors."

"Did you hear any noises?"

"None at all. The men describe some rather peculiar noises and Sergeant Jervis is an old file and pretty apt to get things straight, but they may have been made by the men who were in trouble. I saw a man caught by a boa in South America once, and the noises he made might very well have been described in almost the same words as Jervis used."

"Thanks, Lieutenant," replied the Doctor. "I'll remember what you have told me. Now I think that we'll go into the cave."

"My orders are to allow no one to enter, Doctor."

"I beg your pardon. Carnes, where is that letter from the Secretary of War?"

Carnes produced the document. The lieutenant examined it and excused himself. He returned in a few moments with the commanding officer.

"In the face of that letter, Dr. Bird," said the major, "I have no alternative to allowing you to enter the cave, but I will warn you that it is at your own peril. I'll give you an escort, if you wish."

"If Lieutenant Pearce will come with me as a guide, that will be all that I need."

The lieutenant paled slightly, but threw back his shoulders.

"Do you wish to start at once, sir?" he asked.

"In a few moments. What is the floor of the cave like where we are going?"

"Quite wet and slimy, sir."

"Very slippery?"

"Yes, sir."

"In that case before we go in we want to put on baseball shoes with cleats on them, so that we can run if we have to. Can you get us anything like that?"

"In a few moments, sir."

"Good! As soon as we can get them we'll start. In the meantime, may I look at that gun that was found?"

The Browning machine-gun was laid before the doctor. He looked it over critically and sniffed delicately at it. He took from his pocket a phial of liquid, moistened a portion of the water-jacket of the weapon, and then rubbed the moistened part briskly with his hand. He sniffed again. He looked disappointed, and again examined the gun closely.

"Carnes," he said at length, "do you see anything on this gun that looks like tooth marks?"

"Nothing, Doctor."

"Neither do I. There are some marks here which might quite conceivably be finger-prints of a forty-foot giant, and those two parallel grooves look like the result of severe squeezing, but there are no tooth marks. Strange. There is no persistent odor on the gun, which is also strange. Well, there's no use in theorizing: we are confronted by a condition and not a theory, as someone once said. Let's put on those baseball shoes and see what we can find out."

Dr. Bird led the way into the cave, Carnes and the lieutenant following closely with electric torches. In each hand Dr. Bird carried a phosphorus hand-grenade. No other weapons were visible, although the doctor knew that Carnes carried a caliber .45 automatic pistol strapped under his left armpit. As they passed into the cave the lieutenant stepped forward to lead the way.

"I'm going first," said the doctor. "Follow me and indicate the turns by pressure on my shoulder. Don't speak after we have started, and be ready for instant flight. Let's go."

Forward into the interior of the cave they made their way. The iron cleats of the baseball shoes rang on the floor and the noise echoed back and forth between the walls, dying out in little eerie whispers of sound that made Carnes' hair rise. Ever forward they pressed, the lieutenant guiding the doctor by silent pressure on his shoulder and Carnes following closely. For half a mile they went on until a restrainable pressure brought the doctor to a halt. The lieutenant pointed silently toward a crack in the wall before them. Carnes started forward to examine it, but a warning gesture from the doctor stopped him.

Slowly, an inch at a time, the doctor crept forward, hand-grenades in readiness. Presently he reached the crack and, shifting one of the grenades into his pocket, he drew forth an electric torch and sent a beam of light through the crack into the dark interior of the earth.

For a moment he stood thus, and then suddenly snapped off his torch and straightened up in an attitude of listening. The straining ears of Carnes and Lieutenant Pearce could hear a faint slithering noise coming toward them, not from the direction of the crack, but from the interior of the cave. Simultaneously a faint, musky, reptilian odor became apparent.

"Run!" shouted the doctor. "Run like hell! It's loose in the cave!"

The lieutenant turned and fled at top speed toward the distant entrance to the cave, Carnes at his heels. Dr. Bird paused for an instant, straining his ears, and then threw a grenade. A blinding flash came from the point where the missile struck and a white cloud rose in the air. The doctor turned and fled after his companions. Not for nothing had Dr. Bird been an athlete of note in his college days. Despite the best efforts of his companions, who were literally running for their lives, he soon caught up with them. As he did so a weird, blood-curdling screech rose from the darkness behind them. Higher and higher in pitch the note rose until it ended suddenly in a gurgling grunt, as though the breath which uttered it had been suddenly cut off. The slithering, rustling noise became louder on their trail.

"Faster!" gasped the doctor, as he put his hand on Carnes' shoulder and pushed him forward.

The noise of pursuit gained slightly on them, and a sound as of intense breathing became audible. Dr. Bird paused and turned and faced the oncoming horror. His electric torch revealed nothing, but he listened for a moment, and then threw his second grenade. Keenly he watched its flight. It flew through the air for thirty yards and then struck an invisible obstruction and bounded toward the ground. Before it struck the downward motion ceased, and it rose in the air. As it rose it burst with a sharp report, and a wild scream of pain filled the cavern with a deafening roar. The doctor fled again after his companions.

By the time he overtook them the entrance of the cave loomed before them. With sobs of relief they burst out into the open. The guards sprang forward with raised rifles, but Dr. Bird waved them back.

"There's nothing after us, men," he panted. "We got chased a little way, but I tossed our pursuer a handful of phosphorus and it must have burned his fingers a little, judging from the racket he made. At any rate, it stopped the pursuit."

The major hurried up.

"Did you see it, Doctor?" he asked.

"No, I didn't. No one has ever seen it or anything like it. I heard it and, from its voice, I think it has a bad cold. At least, it sounded hoarse, so I gave it a little white phosphorus to make a poultice for its throat, but I didn't get a glimpse of it."

"For God's sake, Doctor, what is it?"

"I can't tell you yet, Major. So far I can tell, it is something new to science and I am not sure just what it looks like. However, I hope to be able to show it to you shortly. Is there a telegraph office here?"

"No, but we have a Signal Corps detachment with us, and they have a portable radio set which will put us in touch with the army net."

"Good! Can you place a tent at my disposal?"

"Certainly, Doctor."

"All right, I'll go there, and I would appreciate it if you would send the radio operator to me. I want to send a message to the Bureau of Standards to forward me some apparatus which I need."

"I'll attend to it, Doctor. Have you any special advice to give me about the guarding?"

"Yes. Have you, or can you get, any live stock?"

"Live stock?"

"Yes. Cattle preferred, although hogs or sheep will do at a pinch. Sheep will do quite well."

"I'll see what I can do, Doctor."

"Get them by all means, if it is possible to do so. Don't worry about paying for them: secret service funds are not subject to the same audit that army funds get. If you can locate them, drive a couple of cattle or half a dozen sheep well into the cave and tether them there. If you don't get them, have your sentries posted well away from the cave mouth, and if any disturbance occurs during the night, tell them to break and run. I hope it won't come out, but I can't tell."

A herd of cattle was soon located and two of the beasts driven into the cave. Two hours later a series of horrible screams and bellowings were heard in the cave. Following their orders the sentries abandoned their posts and scattered, but the noise came no nearer the mouth, and in a few minutes silence again reigned.

"I hope that will be all that will be needed for a couple of days," said the doctor to the commanding officer, "but you had better have a couple more cattle driven in in the morning. We want to keep the brute well fed. Is there a tank stationed at Fort Thomas?"

"No, there isn't."

"Then radio Washington that I want the fastest three-man tank that the army has sent here at once. Don't bother with military channels, radio direct to the Adjutant General, quoting the Secretary of the Treasury as authority. Tell him that it's a rush matter, and sign the message 'Bird' if you are afraid of getting your tail twisted."

Twice more before the apparatus which the doctor had ordered from Washington arrived cattle were driven into the depths of the cave, and twice were the screams and bellowings from the cave repeated. Each time searching parties found the cattle gone in the morning. A week after the doctor's arrival, a special train came up, carrying four mechanics from the Bureau of Standards, together with a dozen huge packing cases. Under the direction of the doctor the cases were unpacked and the apparatus put together. Before the assembly had been completed the tank which had been requested arrived from Camp Meade, and the Bureau mechanics began to install some of the assembled units in it.

The first apparatus which was installed in the tank consisted of an electric generator of peculiar design which was geared to the tank motor. The electromotive force thus generated was led across a spark gap with points of a metallic substance. The light produced was concentrated by a series of parabolic reflectors, directed against a large quartz prism, and thence through a lens which was designed to throw a slightly divergent beam.

"This apparatus," Dr. Bird explained to the Signal Corps officer, who was an interested observer, "is one which was designed at the Bureau for the large scale production of ultra-violet light. There is nothing special about the generator except that it is highly efficient and gives an almost constant electromotive force. The current thus produced is led across these points, which are composed of magnalloy, a development of the Bureau. We found on investigation that a spark gave out a light which was peculiarly rich in ultra-violet rays when it was passed between magnesium points. However, such points could not be used for the handling of a steady current because of lack of durability and ease of fusion, so a mixture of graphite, alundum and metallic magnesium was pressed together with a binder which will stand the heat. Thus we get the triple advantages of ultra-violet light production, durability, and high resistance.

"The system of reflectors catches all of the light thus produced except the relatively small portion which goes initially in the right direction, and directs it on this quartz prism where, due to the refractive powers of the prism, the light is broken up into its component parts. The infra-red rays and that portion of the spectrum which lies in the visible range, that is, from red to violet inclusive, are absorbed by a black body, leaving only the ultra-violet portion free to send a beam through this quartz lens."

"I thought that a lens would absorb ultra-violet light," objected the signal officer.

"A lens made of glass will, but this lens is made of rock crystal, which is readily permeable to ultra-violet. The net result of this apparatus is that we can direct before us as we move in the tank a beam of light which is composed solely of the ultra-violet portion of the spectrum."

"In other words, an invisible light?"

"Yes. That is, invisible to the human eye. The effect of this beam of ultra-violet light in the form of severe sunburn would be readily apparent if you exposed your skin to it for any length of time, and the effects on your eyesight of continued gazing would be apt to be disastrous. It would produce a severe opthalmia and temporary impairment of the vision, somewhat the same symptoms as are observed in snow blindness."

"I see. May I ask what is the object of the whole thing?"

"Surely. Before we can successfully combat this peculiar visitant from another world, it is necessary that we gain some idea of the size and appearance of it. Nothing of the sort has before made its appearance, so far as the annals of science go, and so I am forced to make some rather wild guesses at the nature of the animal. You are probably aware of the fact that the property of penetration possessed by all waves is a function of their frequency, or, perhaps I should say, of their wave-length?"

"Certainly."

"The longer rays of visible light will not penetrate as deeply into a given substance as the shorter ultra-violet rays. This visitor is evidently from some unexplored and, indeed, unknown cavern in the depths of the earth where visible light has never penetrated. Apparently in this cavern the color of the inhabitants is ultra-violet, and hence invisible to us."

"You are beyond my depth, Doctor."

"Pardon me. You understand, of course, what color is? When sunlight, which is a mixture of all colors from infra-red to ultra-violet inclusive, falls on an object, certain rays are reflected and certain others are absorbed. If the red rays are reflected and all others absorbed, the object appears red to our eyes. If all the rays are reflected, the object appears white, and if all are absorbed, it appears black."

"I understand that."

"The human eye cannot detect ultra-violet. Suppose then, that we have an object, either animate or inanimate, the surface of which reflects only ultra-violet light, what will be the result? The object will be invisible."

"I should think it would be black if all the rays except the ultra-violet were absorbed."

"It would, but mark, I did not say the others were absorbed. Are you familiar with fluorescein?"

"No."

"I think you are. It is the dye used in making changeable silk. If we fill a glass container with a fluorescein solution and look at it by reflected light it appears green. If we look at it by transmitted light, that is, light which has traversed the solution, it appears red. In other words, this is a substance which reflects green light, allows a free passage to red light, and absorbs all other light. This creature we are after, if my theory is correct, is composed of a substance which allows free passage to all of the visible light rays and at the same time reflects ultra-violet light. Do I make this clear?"

"Perfectly."

"Very well, then. My apparatus will project forward a beam of ultra-violet light which will be in much greater concentration than exists in an incandescent electric light. It is my hope that this light will be reflected by the body of the creature to a sufficient to allow me to make a photograph of it."

"But won't your lens prevent the ultra-violet light from reaching your plate?"

"An ordinary lens made of optical glass would do so, but I have a camera here equipped with a rock crystal lens, which will allow ultra-violet light to pass through it practically unhindered, and with very slight distortion. When I add that I will have my camera charged with X-ray film, a film which is peculiarly sensitive to the shorter wave-lengths, you will see that I will have a fair chance of success."

"It sounds logical. Would you allow me to accompany you when you make your attempt?"

"I will be glad of your company, if you can drive a tank. I want to take Carnes with me, and the tank will only hold two besides the driver."

"I can drive a tractor."

"In that case you should master the tricks of tank driving in short order. Get familiar with it and we'll appoint you as driver. We'll be ready to go in to-night, but I am going to wait a day. Our friend was fed last night, and there is less chance he'll be about."

The early part of the next evening was marked by howls and screams coming from the mouth of the cave. As the night wore on the noises were quite evidently coming nearer and the sentries watched the cave mouth nervously, ready to bolt and scatter according to their orders at the first alarm. About two A. M. the doctor and Carnes climbed into the tank beside Lieutenant Leffingwell, and the machine moved slowly into the cave. A search-light on the front of the tank lighted the way for them and, attached to a frame which held it some distance ahead of them, was a luckless sheep.

"Keep your eye on the mutton, Carnes," cautioned the doctor. "As soon as anything happens to it, shut off the search-light and let me try to get a picture. As soon as I have made my exposures I'll tell you, and you can snap it on again. Lieutenant, when the picture is made, turn your tank and make for the entrance to the cave. If we are lucky, we'll get out."

Forward the tank crawled, the sheep bleating and trying to break loose from the bonds which held it. It was impossible to hear much over the roar of the motor, but presently Dr. Bird leaned forward, his eyes shining.

"I smell musk," he announced. "Get ready for action."

Even as he spoke the sheep was suddenly lifted into the air. It gave a final bleat of terror, and then its head was torn from its body.

"Quick, Carnes!" shouted the doctor.

The search-light went out, and Carnes and the lieutenant could hear the slide of the ultra-violet light which Dr. Bird was manipulating open. For two or three minutes the doctor worked with his apparatus.

"All right!" he cried suddenly. "Lights on and get out of here!"

Carnes snapped on the search-light and Lieutenant Leffingwell swung the tank around and headed for the cave mouth. For a few feet their progress was unhindered and then the tank ceased its forward motion, although the motor still roared and the track slid on the cave floor. Carnes watched with horror as one side of the tank bent slowly in toward him. There was a rending sound, and a portion of the heavy steel fabric was torn away. Dr. Bird bent over something on the floor of the tank. Presently he straightened up and threw a small object into the darkness. There was a flash of light, and bits of flaming phosphorus flew in every direction. The anchor which held the tank was suddenly loosed and the machine crawled forward at full speed, while a roar as of escaping air mingled with a bellowing shriek burdened the smoke-laden air.

"Faster!" cried the doctor, as he threw another grenade.

Lieutenant Leffingwell got the last bit of speed possible out of the tank and they reached the cave mouth without further molestation.

"I had an idea that our friend wouldn't care to pass through a phosphorus screen," said Dr. Bird with a chuckle as he climbed out of the tank. "He must have been rather severely burned the other day, and once burned is usually twice shy. Where is Major Brown?"

The commanding officer stepped forward.

"Drive a couple of cattle into the cave, Major," directed Dr. Bird. "I want to fill that brute up and keep him quiet for a while. I'm going to develop my films."

Lieutenant Leffingwell and Carnes peered over the doctor's shoulders as he manipulated his films in a developing bath. Gradually vague lines and blotches made their appearance on one of the films, but the form was indistinct. Dr. Bird dropped the films in a fixing tank and straightened up.

"We have something, gentlemen," he announced, "but I can't tell yet how clear it is. It will take those films fifteen minutes to fix, and then we'll know."

In a quarter of an hour he lifted the first film from the tank and held it to the light. The film showed a blank. With an exclamation of disappointment he lifted a second and third film from the tank, with the same result He raised the fourth one.

"Good Lord!" gasped Carnes.

In the plate could be plainly seen the hind quarters of the sheep held in the grasp of such a monster as even the drug-laden brain of an opium smoker never pictured. Judging from the sheep, the monster stood about twenty feet tall, and its frame was surmounted by a head resembling an overgrown frog. Enormous jaws were opened to seize the sheep but, to the amazement of the three observers, the jaws were entirely toothless. Where teeth were to be expected, long parallel ridges of what looked like bare bone, appeared, without even a rudimentary segregation into teeth. The body of the monster was long and snakelike, and was borne on long, heavy legs ending in feet with three long toes, armed with vicious claws. The crowning horror of the creature was its forelegs. There were of enormous length, thin and attenuated looking, and ended in huge misshapen hands, knobby and blotched, which grasped the sheep in the same manner as human hands. The eyes were as large as dinner plates, and they were glaring at the camera with an expression of fiendish malevolence which made Carnes shudder.

"How does that huge thing ever get through that crack we examined?" demanded the lieutenant.

Dr. Bird rubbed his head thoughtfully.

"It's not an amphibian," he muttered, "as is plainly shown by the shape of the limbs and the lack of a tail, and yet it appears to have scales of the true fish type. It corresponds to no recovered fossil, and I am inclined to believe it is unique. The nervous organisation must be very low, judging from the lack of forehead and the general conformation. It has enormous strength, and yet the arms look feeble."

"It can't get through that crack," insisted the lieutenant.

"Apparently not," replied the doctor. "Wait a moment, though. Look at this!"

He pointed to the great disproportion between the length and diameter of the forelegs, and then to the hind legs.

"Either this is grave distortion or there is something mighty queer about that conformation. No animal could be constructed like that."

He turned the film so that an oblique light fell on it. As he did so he gave a cry of astonishment.

"Look here!" he said sharply. "It does get through that crack! Look at those arms and hands! There is the answer. This creature is tall and broad, but from front to rear it can measure only a few inches. The same must be true of the froglike head. That animal has been developed to live and move in a low roofed cavern, and to pass through openings only a few inches wide. Its bulk is all in two dimensions!"

"I believe you're right," said Carnes as he studied the film.

"There is no doubt of it," answered the doctor. "Look at those paws, too, Carnes. That substance isn't bone, it's gum. The thing is so young and helpless that it hasn't cut its teeth yet. It must be a baby, and that is the reason why it made its way into the cave when no other of its kind ever has."

"How large are full grown ones if this is a baby?" asked the lieutenant.

"The Lord alone knows," replied Dr. Bird. "I hope that I never have to face one and find out. Well, now that we know what we are fighting, we ought to be able to settle its hash."

"High explosive?" suggested the lieutenant.

"I don't think so. With such a low nervous organization, we would have to tear it practically to pieces to kill it, and I am anxious to keep it from mutilation for scientific study. I have an idea, but I'll have to study a while before I am sure of the details. Send me the radio operator."

The next day the Bureau mechanics began to dismount the apparatus from the tank and to assemble another elaborate contrivance. Before they had made an end of the work additional equipment arrived from Washington, which was incorporated in the new set-up. At length Dr. Bird pronounced himself ready for the attempt.

Under his direction, three cattle were driven into the cave and there tethered. They were there the next morning unharmed, but the second night the now familiar bellowing and howling came from the depths of the cave and in the morning two of the cattle were gone.

"That will keep him quiet for a day or two," said the doctor, "and now to work!"

The tank made its way into the cave, dragging after it two huge cables which led to an engine-driven generator outside the cave. These cables were attached to the terminals of a large motor which was set up in the cave near the place where the cattle were customarily tethered. This motor was the actuating force which turned two generators, one large and one small. The smaller one was mounted on a platform on wheels, which also contained the spark gaps, the reflectors and other apparatus which produced the beam of ultra-violet light which had been used to photograph the monster.

From the larger generator led two copper bars. One of these was connected to a huge copper plate which was laid flat on the floor of the cave. The other led to a platform which was erected on huge porcelain insulators some fifteen feet above the floor. Huge condensers were set up on this platform, and Dr. Bird announced himself in readiness.

A steer was dragged into the cave and up a temporary runway which led to the platform containing the condensers, and there tied with the copper bus bar from the larger generator fastened to three flexible copper straps which led around the animal's body. When this had been completed, everyone except the doctor, Carnes, and Lieutenant Leffingwell left the cave. These three crouched behind the search-light which sent a mild beam of ultra-violet onto the platform where the steer was held. The engine outside the cave was started, and the three men waited with tense nerves.

For several hours nothing happened. The steer tried from time to time to move and, finding it impossible, set up plaintive bellows for liberty.

"I wish something would happen," muttered the lieutenant. "This is getting on my nerves."

"Something is about to happen," replied Dr. Bird grimly. "Listen to that steer."

The bellowing of the steer had suddenly increased in volume and, added to the note of discontent, was a note of fright which had previously been absent. Dr. Bird bent over his ultra-violet search-light and made some adjustments. He handed a helmetlike arrangement to each of his companions and slipped one on over his head.

"I can't see a thing, Doctor," said Carnes in a muffled voice.

"The objects at which you are looking absorb rather than reflect ultra-violet light," said the doctor. "This is a sort of a fluoroscope arrangement, and it isn't perfect at all. However, when the monster comes along, I am pretty sure that you will be able to see it. You may see a little more as your eyes get accustomed to it."

"I can see very dimly," announced the lieutenant in a moment.

Dimly the walls of the cave and the platform before them began to take vague shape. The three stared intently down the beam of ultra-violet light which the doctor directed down the passageway leading deeper into the cave.

"Good Lord!" ejaculated Carnes suddenly.

Slowly into the field of vision came the hideous figure they had seen on the film. As it moved forward a rustling, slithering sound could be heard, even over the bellowing of the steer and the hum of the apparatus. The odor of musk became evident.

Along the floor toward them the thing slid. Presently it reared up on its hind legs and its enormous bulk became evident. It turned somewhat sideways and the correctness of Dr. Bird's hypothesis as to its peculiar shape was proved. All of the bulk of the creature was in two dimensions. Forward it moved, and the horrible human hands stretched forward, while the mouth split in a wide, toothless grin. Nearer the doomed steer the creature approached, and then the reaching hands closed on the animal.

There was a blinding flash, and the monster was hurled backward as though struck by a thunderbolt, while a horrible smell of musk and burned flesh filled the air.

"After it! Quick!" cried the doctor as he sprang forward.

Before he could reach the prostrate creature it moved and then, slowly at first, but with rapidly gaining speed, it slithered over the floor in retreat. Dr. Bird's hand swung through an arc, and there was a deafening crash as a hand-grenade exploded on the back of the fleeing monster.

An unearthly scream came from the creature, and its motion changed from a steady forward glide to a series of convulsive jerks. Leffingwell and Carnes threw grenades, but they went wide of their mark, and the monster began to again increase its speed. Another volley of grenades was thrown and one hit scored, which slowed the monster somewhat but did not arrest the steady forward movement.

"Any more bombs?" demanded the doctor.

"Damn!" he cried as he received negative answers. "The current wasn't strong enough. It's going to get away."

Carnes jerked his automatic from under his armpit and poured a stream of bullets into the fleeing monster. Slower and slower the motion of the creature became, and its movements again became jerky and convulsive.

"Keep it in sight!" cried the doctor. "We may get it yet!"

Cautiously the three men followed the retreating horror, Leffingwell pushing before him the platform holding the ultra-violet ray apparatus. The chase led them over familiar ground.

"There is the crack!" cried the lieutenant.

"Too late!" replied the doctor.

He rushed forward and seized the lower limb of the monster and tried with all his strength to arrest its flight, but despite all that he could do it slid sideways through the crack in the wall and disappeared. A final backward kick of its leg threw the doctor twenty feet against the far wall of the cave.

"Are you hurt, Doctor?" cried Carnes.

"No, I'm all right. Put on your masks and start the gas! Quick! That may stop it before it gets in far!"

The three adjusted gas masks and thrust the mouths of two gas cylinders which were on the light truck into the crack, and opened the valves. The hissing of the gas was accompanied by a thrashing, writhing sound from the bowels of the earth for a few minutes, but the sound retreated and finally died away into an utter silence.

"And that's that!" cried the doctor half an hour later as they took off their gas masks outside the cave. "It got away from us. Carnes, how soon can we get a train back to Washington?"

"What kind of a report are you going to make to the Bureau, Doctor?" asked Carnes as they sat in the smoker of a southern train, headed for the capital.

"I'm not going to put in any report, Carnes," replied the doctor. "I haven't got the creature or any part of it to show, and no one would believe me. I am going to maintain a discreet silence about the whole matter."

"But you have your photograph to show, Doctor, and you have my evidence and Lieutenant Leffingwell's."

"The photograph might have been faked and I might have doped both of you. In any case, your words are no better than mine. No, indeed, Carnes, when I failed to make the current strong enough to kill it outright I made the first of the moves which bind me to silence, although I thought that two hundred thousand volts would be enough.

"The second failure I made was when I missed him with my second grenade, although I doubt if all six would have stopped him. My third failure was when we failed to get a sufficient concentration of cyanide gas into that hole in a hurry. The thing is so badly crippled that it will die, but it may take hours, or even days, for it to do so. It has already made its way so far into the earth that we couldn't reach it by blasting without danger of bringing the whole place down on our heads. Even if we could blast our way into the place it came from I wouldn't dare open a path which would allow Lord only knows what terrible monsters to invade the earth. When the soldiers have finished stopping that crack with ten feet of solid masonry, I think the barrier will hold, even against that critter's papa and mamma and all its relatives. Then Mammoth Cave will be safe for visitors again. That latter fact is the only report which I will make."

"It is a dandy story to go to waste," said Carnes soberly.

"Tell it then, if you wish, and get laughed at for your pains. No, Carnes, you must learn one thing. A man like Bolton, for instance, will implicitly believe that a four leaf clover in his watch-charm will bring him good luck, and that carrying a buckeye keeps rheumatism away from him; but tell him a bit of sober fact like this, attested by three reliable witnesses and a good photograph, and you'll just get laughed at for your pains. I'm going to keep my mouth shut."

"So be it, then!" replied Carnes with a sigh.

PHANTOMS OF REALITY

A COMPLETE NOVEL

By Ray Cummings

CHAPTER I
WALL STREET—OR THE OPEN ROAD?

Red Sensua's knife came up dripping—and the two adventurers knew that chaos and bloody revolution had been unleashed in that shadowy kingdom of the fourth dimension.

When I was some fifteen years old, I once made the remark, "Why, that's impossible."

The man to whom I spoke was a scientist. He replied gently, "My boy, when you are grown older and wiser you will realize that nothing is impossible."

Somehow, that statement stayed with me. In our swift-moving wonderful world I have seen it proven many times. They once thought it impossible to tell what lay across the broad, unknown Atlantic Ocean. They thought the vault of the heavens revolved around the earth. It was impossible for it to do anything else, because they could see it revolve. It was impossible, too, for anything to be alive and yet be so small that one might not see it. But the microscope proved the contrary. Or again, to talk beyond the normal range of the human voice was impossible, until the telephone came to show how simply and easily it might be done.

I never forgot that physician's remark. And it was repeated to me some ten years later by my friend, Captain Derek Mason, on that memorable June night of 1929.

My name is Charles Wilson. I was twenty-five that June of 1929. Although I had lived all of my adult life in New York City, I had no relatives there and few friends.

I had known Captain Mason for several years. Like myself, he seemed one who walked alone in life. He was an English gentleman, perhaps thirty years old. He had been stationed in the Bermudas, I understood, though he seldom spoke of it.

I always felt that I had never seen so attractive a figure of a man as this Derek Mason. An English aristocrat, he was, straight and tall and dark, and rather rakish, with a military swagger. He affected a small, black mustache. A handsome, debonair fellow, with an easy grace of manner: a modern d'Artagnan. In an earlier, less civilized age, he would have been expert with sword and stick, I could not doubt. A man who could capture the hearts of women with a look. He had always been to me a romantic figure, and a mystery that seemed to shroud him made him no less so.

A friendship had sprung up between Derek Mason and me, perhaps because we were such opposite types! I am an American, of medium height, and medium build. Ruddy, with sandy hair. Derek Mason was as meticulous of his clothes, his swagger uniforms, as the most perfect Beau Brummel. Not so myself. I am careless of dress and speech.

I had not seen Derek Mason for at least a month when, one June afternoon, a note came from him. I went to his apartment at eight o'clock the same evening. Even about his home there seemed a mystery. He lived alone with one man servant. He had taken quarters in a high-class bachelor apartment building near lower Fifth Avenue, at the edge of Greenwich Village.

All of which no doubt was rational enough, but in this building he had chosen the lower apartment at the ground-floor level. It adjoined the cellar. It was built for the janitor, but Derek had taken it and fixed it up in luxurious fashion. Near it, in a corner of the cellar, he had boarded off a square space into a room. I understood vaguely that it was a chemical laboratory. He had never discussed it, nor had I ever been shown inside it. Unusual, mysterious enough, and that a captain of the British military should be an experimental scientist was even more unusual. Yet I had always believed that for a year or two Derek had been engaged in some sort of chemical or physical experiment. With all his military swagger he had the precise, careful mode of thought characteristic of the man of scientific mind.

I recall that when I got his note with its few sentences bidding me come to see him, I had a premonition that it marked the beginning of something strange. As though the portals of a mystery were opening to me!

Nothing is impossible! Nevertheless I record these events into which I was plunged that June evening with a very natural reluctance. I expect no credibility. If this were the year 2000, my narrative doubtless would be tame enough. Yet in 1929 it can only be called a fantasy. Let it go at that. The fantasy of to-day is the sober truth of to-morrow. And by the day after, it is a mere platitude. Our world moves swiftly.

Derek received me in his living-room. He admitted me himself. He told me that his man servant was out. It was a small room, with leather-covered easy chairs, rugs on its hardwood floor, and sober brown portieres at its door and windows. A brown parchment shade shrouded the electrolier on the table. It was the only light in the room. It cast its mellow sheen upon Derek's lean graceful figure as he flung himself down and produced cigarettes.

He said, "Charlie, I want a little talk with you. I've something to tell you—something to offer you."

He held his lighter out to me, with its tiny blue alcohol flame under my cigarette. And I saw that his hand was trembling.

"But I don't understand what you mean," I protested.

He retorted, "I'm suggesting that you might be tired of being a clerk in a brokerage office. Tired of this humdrum world that we call civilization. Tired of Wall Street."

"I am, Derek. Heavens, that's true enough."

His eyes held me. He was smiling half whimsically: his voice was only half serious. Yet I could see, in the smoldering depths of those luminous dark eyes, a deadly seriousness that belied his smiling lips and his gay tone.

He interrupted me with, "And I offer you a chance for deeds of high adventuring. The romance of danger, of pitting your wits against villainy to make right triumph over wrong, and to win for yourself power and riches— and perhaps a fair lady...."

"Derek, you talk like a swashbuckler of the middle ages."

I thought he would grin, but he turned suddenly solemn.

"I'm offering to make you henchman to a king, Charlie."

"King of what? Where?"

He spread his lean brown hands with a gesture. He shrugged. "What matter? If you seek adventure, you can find it—somewhere. If you feel the lure of romance—it will come to you."

I said, "Henchman to a king?"

But still he would not smile. "Yes. If I were king. I'm serious. Absolutely. In all this world there is no one who cares a damn about me. Not in this world, but...."

He checked himself. He went on, "You are the same. You have no relatives?"

"No. None that ever think of me."

"Nor a sweetheart. Or have you?"

"No," I smiled. "Not yet. Maybe never."

"But you are too interested in Wall Street to leave it for the open road?" He was sarcastic now. "Or do you fear deeds of daring? Do you want to right a great wrong? Rescue an oppressed people, overturn the tyranny of an evil monarch, and put your friend and the girl he loves upon the throne? Or do you want to go down to work as usual in the subway to-morrow morning? Are you afraid that in this process of becoming henchman to a king you may perchance get killed?"

I matched his caustic tone. "Let's hear it, Derek."

CHAPTER II
THE CHALLENGE OF THE UNKNOWN

Incredible! Impossible! I did not say it, though my thoughts were written on my face, no doubt.

Derek said quietly, "Difficult to believe, Charlie? Yes! But it happens to be true. The girl I love is not of this world, but she lives nevertheless. I have seen her, talked with her. A slim little thing—beautiful...."

He sat staring. "This is nothing supernatural, Charlie. Only the ignorant savages of our past called the unknown—the unusual—supernatural. We know better now."

I said, "This girl—"

He gestured. "As I told you, I have for years been working on the theory that there is another world, existing here in this same space with us. The Fourth Dimension! Call it that it you like. I have found it, proved its existence! And this girl—her name is Hope—lives in it. Let me tell you about her and her people. Shall I?"

My heart was pounding so that it almost smothered me. "Yes, Derek."

"She lives here, in this Space we call New York City. She and her people use this same Space at the same time that we use it. A different world from ours, existing here now with us! Unseen by us. And we are unseen by them!

"A different form of matter, Charlie. As tangible to the people of the other realm as we are to our own world. Humans like ourselves."

He paused, but I could find no words to fill the gap. And presently he went on:

"Hope's world, co-existing here with us, is dependent upon us. They speak what we call English. They shadow us."

I murmured, "Phantoms of reality."

"Yes. A world very like ours. But primitive, where ours is civilized."

He paused again. His eyes were staring past me as though he could see through the walls of the cellar room into great reaches of the unknown.

What a strange mixture was this Derek Mason! What a strange compound of the cold reality of the scientist and the fancy of the romantic dreamer! Yet I wonder if that is not what science is. There is no romantic lover gawping at the moon who could have more romance in his soul, or see in the moonlit eyes of his loved one more romance than the scientist finds in the wonders of his laboratory.

Derek went on slowly:

"A primitive world, primitive nation, primitive passions! As I see it now, Charlie—as I know it to be—it seems as though perhaps Hope's world is merely a replica of ours, stripped to the primitive. As though it might be the naked soul of our modern New York, ourselves as we really are, not as we pretend to be."

He roused himself from his reverie.

"Hope's nation is ruled by a king. An emperor, if you like. A monarch, beset with the evils of luxury and ease, and wine and women. He is surrounded by his nobles, the idle aristocracy, by virtue of their birth proclaiming themselves of too fine a clay to work. The crimson nobles, they are called. Because they affect crimson cloaks, and their beautiful women, voluptuous, sex-mad, are wont to bedeck themselves in veils and robes of crimson.

"And there are workers, toilers they call them. Oppressed, down-trodden toilers, with hate for the nobles and the king smoldering within them. In France there was such a condition, and the bloody revolution came of it. It exists here now. Hope was born in the ranks of these toilers, but has risen by her grace and beauty to a position in the court of this graceless monarch."

He leaped from his chair and began pacing the room. I sat silent, staring at him. So strange a thing! Impossible? I could not say that. I could only say, incredible to me. And as I framed the thought I knew its incredibility was the very measure of my limited intelligence, my lack of knowledge. The vast unknown of nature, so vast that everything which was real to me, understandable to me, was a mere drop in the ocean of the existing unknown.

"Don't you understand me now?" Derek added vehemently. "I'm not talking fantasy. Cold reality! I've found a way to transport myself—and you—into this different state of matter, into this other world! I've already made a test. I went there and stayed just for a few moments, a night or so ago."

It made my heart leap wildly. He went on:—

"There is chaos there. Smoldering revolution which at any time—tonight perhaps—may burst into conflagration and destroy this wanton ruling class." He laughed harshly. "In Hope's world the workers are a primitive, ignorant people. Superstitious. Like the peons of Mexico, they're all primed and ready to shout for any leader who sets himself up. My chance—our chance—"

He suddenly stopped his pacing and stood before me. "Don't you feel the lure of it? The open road? 'The road is straight before me and the Red Gods call for me!' I'm going, Charlie. Going to-night—and I want you to go with me! Will you?"

Would I go? The thing leaped like a menacing shadow risen solidly to confront me. Would I go?

Suddenly there was before me the face of a girl. White. Apprehensive. It seemed almost pleading. A face beautiful, with a mouth of parted red lips. A face framed in long, pale-golden hair with big staring blue eyes. Wistful eyes, wan with starlight—eyes that seemed to plead.

I thought, "Why, this is madness!" I was not seeing this face with my eyes. There was nothing, no one here in the room with me but Derek. I knew it. The shadows about us were empty. I was conjuring the face only from Derek's words, making real that which existed only in my imagination.

Yet I knew that in another realm, with my thoughts now bridging the gap, the girl was real. Would I go into the unknown?

The quest of the unknown. The gauntlet of the unknown flung down now before me, as it was flung down before the ancient explorers who picked up its challenge and mounted the swaying decks of their little galleons and said, "We'll go and see what lies off there in the unknown."

That same lure was on me now. I heard my voice saying, "Why yes, I guess I'll go, Derek."

CHAPTER III
INTO THE UNKNOWN

We stood in the boarded room which was Derek's laboratory. Our preparations had been simple: Derek had made them all in advance. There was little left to do. The laboratory was a small room of board walls, board ceiling and floor. Windowless, with a single door opening into the cellar of the apartment house.

Derek had locked the door after us as we entered. He said, "I have sent my man servant away for a week. The people in the house here think I have gone away on a vacation. No one will miss us, Charlie—not for a time, anyway."

No one would miss me, save my employers, and to them I would no doubt be small loss.

We had put out the light in Derek's apartment and locked it carefully after us. This journey! I own that I was trembling, and frightened. Yet a strange eagerness was on me.

The cellar room was comfortably furnished. Rugs were on its floor. Whatever apparatus of a research laboratory had been here was removed now. But the evidence of it remained—Derek's long search for this secret which now he was about to use. A row of board shelves at one side of the room showed where bottles and chemical apparatus had stood. A box of electrical tools and odds and ends of wire still lay discarded in a corner of the room. There was a tank of running water, and gas connections, where no doubt bunsen burners had been.

Derek produced his apparatus. I sat on a small low couch against the wall and watched him as he stripped himself of his clothes. Around his waist he adjusted a wide, flat, wire-woven belt. A small box was fastened to it in the middle of the back—a wide, flat thing of metal, a quarter of an inch thick, and curved to fit his body. It was a storage battery of the vibratory current he was using. From the battery, tiny threads of wire ran up his back to a wire necklace flat against his throat. Other wires extended down his arms to the wrists. Still others down his legs to the ankles. A flat electrode was connected to the top of his head like a helmet. I was reminded as he stood there, of medical charts of the human body with the arterial system outlined. But when he dressed again and put on his jaunty captain's uniform, only the electrode clamped to his head and the thin wires dangling from it in the back were visible to disclose that there was anything unusual about him.

He said smilingly, "Don't stare at me like that."

I took a grip on myself. This thing was frightening, now that I actually was embarked on it. Derek had explained to me briefly the workings of his apparatus. A vibratory electronic current, for which as yet he had no name, was stored in the small battery. He had said:

"There's nothing incomprehensible about this, Charlie. It's merely a changing of the vibration rate of the basic substance out of which our bodies are made. Vibration is the governing factor of all states of matter. In its essence what we call substance is wholly intangible. That is already proven. A vortex! A whirlpool of nothingness! It creates a pseudo-substance which is the only material in the universe. And from this, by vibration, is built the complicated structure of things as we see and feel them to be, all dependent upon vibration. Everything is altered, directly as the vibratory rate is changed. From the most tenuous gas, to fluids to solids—throughout all the different states of matter the only fundamental difference is the rate of vibration."

I understood the basic principle of this that he was explaining—that now when this electronic current which he had captured and controlled was applied to our physical body, the vibration rate of every smallest and most minute particle of our physical being was altered. There is so little in the vast scale of natural phenomena of which our human senses are cognisant! Our eyes see the colors of the spectrum, from red to violet. But a vast invisible world of color lies below the red of the rainbow! Physicists call it the infra-red. And beyond the violet, another realm—the ultra-violet. With sound it is the same. Our audible range of sound is very small. There are sounds with too slow a vibratory rate for us to hear, and others too rapid. The differing vibratory rate from most tenuous gas to most substantial solid is all that we can perceive in this physical world of ours. Yet of the whole, it is so very little! This other realm to which we were now going lay in the higher, more rapid vibratory scale. To us, by comparison, a more tenuous world, a shadow realm.

I listened to Derek's words, but my mind was on the practicality of what lay ahead. An explorer, standing upon his ship, may watch his men bending the sails, raising the anchor, but his mind flings out to the journey's end....

We were soon ready. Derek wore his jaunty uniform, I wore my ordinary business suit. A magnetic field would be about us, so that in the transition anything in fairly close contact with our bodies was affected by the current.

Derek said, "I will go first, Charlie."

"But, Derek—" A fear, greater than the trembling I had felt before, leaped at me. Left here alone, with no one on whom to depend!

He spoke with careful casualness, but his eyes were burning me. "Just sit there, and watch. When I am gone, turn on the current as I showed you and come after me. I'll wait for you."

"Where?" I stammered.

He smiled faintly. "Here. Right here. I'm not going away! Not going to move. I'll be here on the couch waiting for you."

Terrifying words! He had lowered the couch, bending out its short legs until the frame of it rested on the board floor. He drew a chair up before it and seated me. He sat down on the couch.

He said, "Oh, one other thing. Just before you start, put out the light. We can't tell how long it will be before we return."

Terrifying words!

His right hand was on his left wrist where the tiny switch was placed. He smiled again. "Good luck to us, Charlie!"

Good luck to us! The open road, the unknown!

I sat there staring. He was partly in shadow. The room was very silent. Derek lay propped up on one elbow. His hand threw the tiny switch.

There was a breathless moment. Derek's face was set and white, but no whiter than my own, I was sure. His eyes were fixed on me. I saw him suddenly quiver and twitch a little.

I murmured, "Derek—"

At once he spoke, to reassure me. "I'm all right, Charlie. That was just the first feel of it."

There was a faint quivering throb in the room, like a tiny distant dynamo throbbing. The current was surging over Derek; his legs twitched.

A moment. The faint throbbing intensified. No louder, but rapid, infinitely more rapid. A tiny throb, an aerial whine, faint as the whirring wings of a humming bird. It went up the scale, ascending in pitch, until presently it was screaming with an aerial microscopic voice.

But there seemed no change in Derek. His uniform was glowing a trifle, that was all. His face was composed now; he smiled, but did not speak. His eyes roved away from me, as though now he were seeing things that I could not see.

Another moment. No change.

Why, what was this? I blinked, gasped. There was a change! My gaze was fastened upon Derek's white face. White? It was more than white now! A silver sheen seemed to be coming to his skin!

I think no more than a minute had passed. His face was glowing, shimmering. A transparent look was coming to it, a thinness, a sudden unsubstantiality! He dropped his elbow and lay on the couch, stretched at full length at my feet. His eyes were staring.

And suddenly I realized that the face that held those staring eyes was erased! A shimmering apparition of Derek was stretched here before me. I could see through it now! Beneath the shimmering, blurred outlines of his body I could see the solid folds of the couch cover. A ghost of Derek here. An apparition—fading—dissipating!

A gossamer outline of him, imponderable, intangible.

I leaped to my feet, staring down over him.

"Derek!"

The shape of him did not move. Every instant it was more vaporous, more unreal.

I thought, "He's gone!"

No! He was still there. A white mist of his form on the couch. Melting, dissipating in the light like a fog before sunshine. A wisp of it left, like a breath, and then there was nothing.

I sat on the couch. I had put out the light. Around me the room was black. My fingers found the small switch at my wrist. I pressed it across its tiny arc.

The first shock was slight, but infinitely strange. A shuddering, twitching sensation ran all over me. It made my head reel, swept a wave of nausea over me, a giddiness, a feeling that I was falling through darkness. I lay on the couch, bracing myself. The current was whining up its tiny scale. I could feel it now. A tiny throbbing, communicating itself to my physical being.

And then in a moment I realized that my body was throbbing. The vibration of the current was communicating itself to the most minute cells of my body. An indescribable tiny quivering within me. Strange, frightening, sickening at first. But the sickness passed, and in a moment I found it almost pleasant.

I could see nothing. The room was wholly dark. I lay on my side on the couch, my eyes staring into the blackness around me. I could hear the humming of the current, and then it seemed to fade. Abruptly I felt a sense of lightness. My body, lying on the couch, pressed less heavily.

I gripped my arm. I was solid, substantial as before. I touched the couch. It was the couch which was changing, not I! The couch cover queerly seemed to melt under my hand!

The sense of my own lightness grew upon me. A lightness, a freedom, pressed me, as though chains and shackles which all my life had encompassed me were falling away. A wild, queer freedom.

I wondered where Derek was. Had I arrived in the other realm? Was he here? I had no idea how much time had passed: a minute or two, perhaps.

Or was I still in Derek's laboratory? The darkness was as solid, impenetrable as ever. No, not quite dark! I saw something now. A glowing, misty outline around me. Then I saw that it was not the new, unknown realm, but still Derek's room. A shadowy, spectral room, and the light, which dimly illumined it, was from outside.

I lay puzzling, my own situation forgotten for the moment. The light came from overhead, in another room of the apartment house. I stared. Around me now was a dim vista of distance, and vague, blurred, misty outlines of the apartment building above me. The shadowy world I had left now lay bare. There was a moment when I thought I could see far away across a spectral city street. The shadows of the great city were around me. They glowed, and then were gone.

A hand gripped my arm in a solid grip. Derek's voice sounded.

"Are you all right?"

"Yes," I murmured. The couch had faded. I was conscious that I had floated or drifted down a few inches, to a new level. The level of the cellar floor beneath the couch. Cellar floor! It was not that now. Yet there was something solid here, a solid ground, and I was lying upon it, with Derek sitting beside me.

I murmured again, "Yes, I'm all right."

My groping hand felt the ground. It was soil, with a growth of vegetation like a grass sward on it. Were we outdoors? It suddenly seemed so. I could feel soft, warm air on my face and had a sense of open distance around me.

A light was growing, a vague, diffused light, as though day were swiftly coming upon us.

I felt Derek fumbling at my wrist. "That's all, Charlie."

There was a slight shock. Derek was pulling me up beside him. I found myself on my feet, with light around me. I stood wavering, gripping Derek. It was as though I had closed my eyes, and now they were suddenly open. I was aware of daylight, color, and movement. A world of normality here, normal to me now because I was part of it. The realm of the unknown!

CHAPTER IV
"HOPE, I CAME...."

I think I was first conscious of a queer calmness which had settled upon me, as though now I had withdrawn contact with the turmoil of our world! Something was gone, and in its place came a calmness. But that was a mere transition. It had passed in a moment. I stood trembling with eagerness, as I know Derek was trembling.

A radiant effulgence of light was around us, clarifying, growing. There was ground beneath our feet, and sky overhead. A rational landscape, strangely familiar. A physical world like my own, but, it seemed, with a new glory upon it. Nature, calmly serene.

I had thought we were standing in daylight. I saw now it was bright starlight. An evening, such as the evening we had just left in our own world. The starlight showed everything clearly. I could see a fair distance.

We stood at the top of a slight rise. I saw gentle, slightly undulating country. A brook nearby wound through a grove of trees and lost itself. Suddenly, with a shock, I realized how familiar this was! We stood facing what in New York City we call West. The contour of this land was familiar enough for me to identify it. A mile or so ahead lay a river; it shimmered in its valley, with cliffs on its further side. Near at hand the open country was dotted with trees and checkered with round patches of cultivated fields. And there were occasional habitations, low, oval houses of green thatch.

The faint flush of a recent sunset lay upon the landscape, mingled with the starlight. A road—a white ribbon in the starlight—wound over the countryside toward the river. Animals, strange of aspect, were slowly dragging carts. There were distant figures working in the fields.

A city lay ahead of us, set along this nearer bank of the river. A city! It seemed a primitive village. All was primitive, as though here might be some lost Indian tribe of our early ages. The people were picturesque, the field workers garbed in vivid colors. The flat little carts, slow moving, with broad-horned oxen.

This quiet village, drowsing beside the calm-flowing river, seemed all very normal. I could fancy that it was just after sundown of a quiet workday. There was a faint flush of pink upon everything: the glory of the sun just set. And as though to further my fancy, in the village by the river, like an angelus, a faint-toned bell was chiming.

We stood for a moment gazing silently. I felt wholly normal. A warm, pleasant wind fanned my hot face. The sense of lightness was gone. This was normality to me.

Derek murmured, "Hope was to meet me here."

And then we both saw her. She was coming toward us along the road. A slight, girlish figure, clothed in queerly vivid garments: a short jacket of blue cloth with wide-flowing sleeves, knee-length pantaloons of red, with tassels dangling from them, and a wide red sash about her waist. Pale golden hair was piled in a coil upon her head....

She was coming toward us along the edge of the road, from the direction of the city. She was only a few hundred feet from us when we first saw her, coming swiftly, furtively it seemed. A low pike fence bordered the road. She seemed to be shielding herself in the shadows beside it.

We stood waiting in the starlight. The nearest figures in the field and on the road were too far away to notice us. The girl advanced. Her white arm went up in a gesture, and Derek answered. She left the road, crossing the field toward us. As she came closer, I saw how very beautiful she was. A girl of eighteen, perhaps, a fantastic little figure with her vivid garments. The starlight illumined her white face, anxious, apprehensive, but eager.

"Derek!"

He said, "Hope, I came...."

I stood silently watching. Derek's arms went out, and the girl, with a little cry, came running forward and threw herself into them.

CHAPTER V
INTRIGUE

"Am I in time, Hope?"

"Yes, but the festival is to-night. In an hour or two now. Oh Derek, if the king holds this festival, the toilers will revolt. They won't stand it—"

"To-night! It mustn't be held to-night! It doesn't give me time, time to plan."

I stood listening to their vehement, half-whispered words. For a moment or two, absorbed, they ignored me.

"The king will make his choice to-night, Derek. He has announced it. Blanca or Sensua for his queen. And if he chooses the Crimson Sensua—" She stammered, then she went on:

"If he does—there will be bloodshed. The toilers are waiting, just to learn his choice."

Derek exclaimed, "But to-night is too soon! I've got to plan. Hope, where does Rohbar stand in this?"

Strange intrigue! I pieced it together now, from their words, and from what presently they briefly told me. A festival was about to be held, an orgy of feasting and merrymaking, of music and dancing. And during it, this young King Leonto was to choose his queen. There were two possibilities. The Crimson Sensua, a profligate, debauched woman who, as queen, would further oppress the workers. And Blanca, a white beauty, risen from the toilers to be a favorite at the Court. Hope was her handmaiden.

If Blanca were chosen, the toilers would be appeased. She was one of them. She would lead this king from his profligate ways, would win from him justice for the workers.

But Derek and Hope both knew that the pure and gentle Blanca would never be the king's choice. And to-night the toilers would definitely know it, and the smoldering revolt would burst into flame.

And there was this Rohbar. Derek said, "He is the king's henchman, Charlie."

ne side at a little distance up the river, banked against the water,
road, low building: the palace of the king. About it were broad
, with shrubs and flowers. The whole was surrounded by a high
nce, spiked on top.

 main gate was near at hand; we left our cart. Close to the gate was
 standing alert, a jaunty fellow in leather pantaloons and leather
with a spiked helmet, and in his hand a huge, sharp-pointed lance.
dens of the palace, what we could see of them, seemed empty—none
 favored few might enter here. But as I climbed from the cart, I got
ression that just inside the fence a figure was lurking. It started away
approached the gate. The guard had not seen it—the drab figure of
in what seemed to be dripping garments, as though perhaps he had
in from the water.

d Derek saw him. He muttered, "They are everywhere."

ope led us to the gate. The guard recognized her. At her imperious
e he stood aside. We passed within. I saw the palace now as a long
d structure of timber and stone, with a high tower at the end of one
 The building fronted the river, but here on the garden side there was
ad doorway up an incline, twenty feet up and over a small bridge,
ing what seemed a dry moat. Beyond it, a small platform, then an oval
vay, the main entrance to the building.

Derek and I, shrouded in our crimson cloaks with hoods covering us to
yes, followed Hope into the palace.

I stood here in the starlight, listening to them. This strange primitive realm. There were no modern weapons here. We had brought none. The current used in our transition would have exploded the cartridges of a revolver. I had a dirk which Hope now gave me, and that was all.

Primitive intrigue. I envisaged this chaotic nation, with its toilers ignorant as the oppressed Mexican peons at their worst. Striving to better themselves, yet, not knowing how. Ready to shout for any leader who might with vainglorious words set himself up as a patriot.

This Rohbar, perhaps, was planning to do just that.

And so was Derek! He said, "Hope, if you could persuade the king to postpone the festival—if Blanca would help persuade him—just until to-morrow night...."

"I can try, Derek. But the festival is planned for an hour or two from now."

"Where is the king?"

"In his palace, near the festival gardens."

She gestured to the south. My mind went back to New York City. This hillock, where we were standing in the starlight beside a tree, was in my world about Fifth Avenue and Sixteenth Street. The king's palace—the festival gardens—stood down at the Battery, where the rivers met in the broad water of the harbor.

Derek was saying, "We haven't much time: can you get us to the palace?"

"Yes. I have a cart down there on the road."

"And the cloaks for Charlie and me?"

"Yes."

"Good!" said Derek. "We'll go with you. It's a long chance; he probably won't postpone it. If he does not, we'll be among the audience. And when he chooses the Red Sensua—"

She shuddered, "Oh, Derek—" And I thought I heard her whisper, "Oh, Alexandre—" and I saw his finger go to his lips.

His arm went around her. She huddled, small as a child against his tall, muscular body.

He said gently, "Don't be afraid, little Hope."

His face was grim, his eyes were gleaming. I saw him suddenly as an instinctive military adventurer. An anachronism in our modern New York City. Born in a wrong age. But here in this primitive realm he was at home.

I plucked at him. "How can you—how can we dare plunge into this thing? Hidden with cloaks, yes. But you talk of leading these toilers."

He cast Hope away and confronted me. "I can do it! You'll see, Charlie." He was very strangely smiling. "You'll see. But I don't want to come into the open right away. Not to-night. But if we can only postpone this accursed festival."

We had been talking perhaps five minutes. We were ready now to start away. Derek said:

"Whatever comes, Charlie, I want you to take care of Hope. Guard her for me, will you?"

I said, "Yes, I will try to."

Hope smiled as she held out her hand to me. "I will not be afraid, with Derek's friend."

Her English was of different intonation from our own, but it was her native language, I could not doubt.

I took her cold, slightly trembling hand. "Thank you, Hope."

Her eyes were misty with starlight. Tender eyes, but the tenderness was not for me.

"Yes," I repeated. "You can depend upon me, Derek."

We left the hillock. A food-laden cart came along the road. The driver, a boy vivid in jacket and wide trousers of red and blue, bravely worn but tattered, ran alongside guiding the oxen. When they had passed we followed, and presently we came to the cloaks Hope had hidden. Derek and I donned them. They were long crimson cloaks with hoods.

Hope said, "Many are gathering for the festival shrouded like that. You will not be noticed now."

Further along the road we reached a little eminence. I saw the river ahead of us, and a river behind us. And a few miles to the south, an open spread of water where the rivers joined. Familiar contours! The Hudson River! The East River. And down at the end of the island, New York Harbor.

Hope gestured that way. "The king's palace is there."

We were soon passing occasional houses, primitive thatched dwellings. I saw inside one. Workers were seated over their frugal evening meal. Always the same vivid garments, jaunty but tattered. We passed one old fellow in a field, working late in the starlight. A man bent with age, but still a tiller of the soil. Hope waved to him and he responded, but the look

he gave us as we hurried by shrouded in our cri[...] hostile.

We came to an open cart. It stood by the road [...] coat and spreading horns was fastened to the fence [...] small rollers like wheels. Seats were in it and a v[...] climbed in and rumbled away.

And this starlit road in our own world wa[...] presently passing close to the river's edge. This quie[...] Why, in our world it was massed with docks! Gr[...] funneled, with storied decks lay here! Under this rive[...] passing vehicles! Tubes, with speeding trains crowde[...]

The reality here was so different! Behind us what [...] was strung along the river. Ahead of us also there wei[...] the city of the workers. A bell was tolling. Along all the [...] see the moving yellow spots of lights on the holiday [...] festival. And there were spots of yellow torchlight from [...]

We soon were entering the city streets. Narrow dir[...] with primitive shacks to the sides. Women came to the d[...] our little cart rumbling hastily past. I was conscious of [...] and conscious of the sullen glances of hate which were flu[...] side, here in this squalid, forlorn section where the worke[...]

Along every street now the carts were passing, conver[...] They were filled, most of them, with young men and g[...] costumes. Some of them, like ourselves, were shrouded in[...] The carts occasionally were piled with flowers. As one lar[...] moving faster rumbled by, a girl in it stood up and pelted m[...] She wore a crimson robe, but it had fallen from her shoul[...] glimpse of her face, framed in flowing dark hair, and of eye[...] in them, mocking me, alluring.

We came at last to the end of the island. There seemed to be[...] more people arriving, or here already. The tip of the island had[...] with a broad canopy behind it. Burning torches of wood g[...] yellow, red and blue fire. A throng of gay young people pro[...] walk, watching the arriving boats.

And here, behind the walk at the water's edge, was a gar[...] and lawn, shrubs and beds of tall vivid flowers. Nooks were he[...] lovers, pools of water glinted red and green with the reflected to[...] one of the pools I saw a group of girls bathing, sportive as dolph[...]

CHAPTER VI
THE KING'S HENCHMAN

The long room was bathed in colored lights. There was an ornate tiled floor. Barbaric draperies of heavy fabric shrouded the archways and windows. It was a totally barbaric apartment. It might have been the audience chamber of some fabled Eastern Prince of our early ages. Yet not quite that either. There was a primitive modernity here. I could not define it, could not tell why I felt this strangeness. Perhaps it was the aspect of the people. The room was crowded with men and gay laughing girls in fancy dress costumes. Half of them at least were shrouded in crimson cloaks, but most of the hoods were back. They moved about, laughing and talking, evidently waiting for the time to come for them to go to the festival. We pushed our way through them.

Derek murmured, "Keep your hood up, Charlie."

A girl plucked at me. "Handsome man, let me see." She thrust her painted lips up to mine as though daring me to kiss them. Hope shoved her away. Her parted cloak showed her white, beautiful body with the dark tresses of her hair shrouding it. Exotically lovely she was, with primitive, unrestrained passions—typical of the land in which she lived.

"This way," whispered Hope. "Keep close together. Do not speak!"

We moved forward and stood quietly against the wall of the room, where great curtains hid us partly from view. Under a canopy, at a table on a raised platform near one end of the apartment, sat the youthful monarch. I saw him as a man of perhaps thirty. He was in holiday garb, robed in silken hose of red and white, a strangely fashioned doublet, and a close-fitting shirt. Bare-headed, with thick black hair, long to the base of his neck.

He sat at the table with a calm dignity. But he relaxed here in the presence of his favored courtiers. He was evidently in a high good humor this night, giving directions for the staging of the spectacle, despatching messengers. I stood gazing at him. A very kingly fellow this. There was about him, that strange mingled look of barbarism and modernity.

Hope approached him and knelt. Derek and I could hear their voices, although the babble of the crowd went on.

"My little Hope, what is it? Stand up, child."

She said, "Your Highness, a message from Blanca."

He laughed. "Say no more! I know it already! She does not want this festival. The workers,"—what a world of sardonic contempt he put into that one word!—"the workers will be offended because we take pleasure to-night. Bah!" But he was still laughing. "Say no more, little Hope. Tell Blanca to dance and sing her best this night. I am making my choice. Did you know that?"

Hope was silent. He repeated, "Did you know that?"

"Yes, Your Highness," she murmured.

"I choose our queen to-night, child. Blanca or Sensua." He sighed. "Both are very beautiful. Do you know which one I am going to choose?"

"No," she said.

"Nor do I, little Hope. Nor do I."

He dismissed her. "Go now. Don't bother me."

She parted her lips as though to make another protest, but his eyes suddenly flashed.

"I would not have you annoy me again. Do you understand?"

She turned away, back toward where Derek and I were lurking. The chattering crowd in the room had paid no attention to Hope, but before she could reach us a man detached himself from a nearby group and accosted her. A commanding figure, he was, I think, quite the largest man in the room. An inch or two taller than Derek, at the least. He wore his red cloak with the hood thrown back upon his wide heavy shoulders. A bullet-head with close-clipped black hair. A man of about the king's age, he had a face of heavy features, and flashing dark eyes. A scoundrel adventurer, this king's henchman.

Hope said, "What is it, Rohbar?"

"You will join our party, little Hope?" He laid a heavy hand on her white arm. His face was turned toward me. I could not miss the gleaming look in his eyes as he regarded her.

"No," she said.

It seemed that he twitched at her, but she broke away from him.

Anger crossed his face, but the desirous look in his eyes remained.

"You are very bold, Hope, to spurn me like this." He had lowered his voice as though fearful that the king might hear him.

"Let me alone!" she said.

She darted away from him, but before she joined us she stood waiting until he turned away.

"No use," Hope whispered. "There is nothing we can do here. You heard what the king said—and the festival is already begun."

Derek stood a moment, lost in thought. He was gazing across the room to where Rohbar was standing with a group of girls. He said at last:

"Come on, Charlie. We'll watch this festival. This damn fool king will choose the Red Sensua." He shrugged. "There will be chaos...."

We shoved our way from the room, went out of the main doorway and hurried through the gardens of the palace. The red-cloaked figures were leaving the building now for the festival grounds. We waited for a group of them to pass so that we might walk alone. As we neared the gate, passing through the shadows of high flowered shrubs, a vague feeling that we were being followed shot through me. In a moment there was so much to see that I forgot it, but I held my hand on my dirk and moved closer to Hope.

We reached the entrance to the canopy. A group of girls, red-cloaked, were just coming out. They rushed past us. They ran, discarding their cloaks. Their white bodies gleamed under the colored lights as they rushed to the pool and dove.

We were just in time. Hope whispered, "The king will be here any moment."

Beneath the canopy was a broad arena of seats. A platform, like a stage, was at one end. It was brilliantly illuminated with colored torches held aloft by girls in flowing robes, each standing like a statue with her light held high. The place was crowded. In the gloom of the darkened auditorium we found seats off to one side, near the open edge of the canopy. We sat, with Hope between us.

Derek whispered, "Shakespeare might have staged a play in a fashion like this."

A primitive theatrical performance. There was no curtain for interlude between what might have been the acts of a vaudeville. The torch girls, like pages, ranged themselves in a line across the front of the stage. They were standing there as we took our seats. The vivid glare of their torches concealed the stage behind them.

There was a few moments wait, then, amid hushed silence, the king with his retinue came in. He sat in a canopied box off to one side. When he was seated, he raised his arm and the buzz of conversation in the audience began again.

Presently the page girls moved aside from the stage. The buzz of the audience was stilted. The performance, destined to end so soon in tragedy, now began.

CHAPTER VII
THE CRIMSON MURDERESS

Hope murmured. "The three-part music comes first. There will first be the spiritual."

An orchestra was seated on the stage in a semi-circle. It was composed of men and women musicians, and there seemed to be over a hundred of them. They sat in three groups; the center group was about to play. In a solemn hush the leaderless choirs, with all its players garbed in white, began its first faint note. I craned to get a clear view of the stage. This white choir seemed almost all wood-wind. There were tiny pipes in little series such as Pan might have used. Flutes, and flageolets; and round-bellied little instruments of clay, like ocarinas. And pitch-pipes, long and slender as a marsh reed.

In a moment I was lost in the music. It began softly, with single muted notes from a single instrument, echoed by the others, running about the choir like a will-o'-the-wisp. It was faint, as though very far away, made more sweet by distance. And then it swelled, came nearer.

I had never heard such music as this. Primitive! It was not that. Nor barbaric! Nothing like the music of our ancient world. Nor was it what I might conceive to be the music of our future. A thing apart, unworldly, ethereal. It swept me, carried me off; it was an exaltation of the spirit lifting me. It was triumphant now. It surged, but there was in its rhythm, the beat of its every instrument, nothing but the soul of purity. And then it shimmered into distance again, faint and exquisite music of a dream. Crooning, pleading, the speech of whispering angels.

It ceased. There was a storm of applause.

I breathed again. Why, this was what music might be in our world but was not. I thought of our blaring jazz.

Hope said, "Now they play the physical music. Then Sensua will dance with Blanca. We will see then which one the king chooses."

On the stage all the torches were extinguished save those which were red. The arena was darker than before. The stage was bathed with a deep crimson. Music of the physical senses! It was, indeed, no more like the other choir than is the body to the spirit.

There were stringed instruments playing now; deep-toned, singing zithers, and instruments of rounded, swelling bodies, like great viols with sensuous, throbbing voices. Music with a swift rhythm, marked by the thump of hollow gourds. It rose with its voluptuous swell into a paean of abandonment, and upon the tide of it, the crimson Sensua flung herself upon the stage. She stood motionless for a moment that all might regard her. The crimson torchlight bathed her, stained crimson the white flush of her limbs, her heavy shoulders, her full, rounded throat.

A woman in her late twenties. Voluptuous of figure, with crimson veils half-hiding, half-revealing it. A face of coarse, sensuous beauty. A face wholly evil, and it seemed to me wholly debauched. Dark eyes with beaded lashes. Heavy lips painted scarlet. A pagan woman of the streets. One might have encountered such a woman swaggering in some ancient street of some ancient city, flaunting the finery given her by a rich and profligate eastern prince.

She stood a moment with smoldering, passion-filled eyes, gazing from beneath her lowered lids. Her glance went to the king's canopy, and flashed a look of confidence, of triumph. The king answered it with a smile. He leaned forward over his railing, watching her intently.

With the surge of the music she moved into her dance. Slowly she began, quite slowly. A posturing and swaying of hips like a nautch girl. She made the rounds of the musicians, leering at them. She stood in the whirl of the music, almost ignoring it, stood at the front of the stage with a gaze of slumberous, insolent passion flung at the king. A knife was in her hand now. She held it aloft. The red torchlight caught its naked blade. With shuddering fancy I seemed to see it dripping crimson. She frowned, and struck it at a phantom lover. She backed away. She stooped and knelt. She knelt and seemed with her empty arms to be caressing a murdered lover's head. She kissed him, rained upon his dead lips her macabre kisses.

And then she was up on her bare feet, again circling the stage. Her anklets clanked as she moved with the tread of a tigress. The musicians shrank from her waving blade.

A girl in white veils was suddenly disclosed standing at the back of the stage.

Derek whispered, "Is that Blanca?"

"Yes," whispered Hope.

Blanca stood watching her rival. The crimson Sensua passed her, took her suddenly by the wrist, drew her forward. For an instant I thought it might have been rehearsed. I saw Blanca as a slim, gentle girl in white, with a white head-dress. A dancer who could symbolize purity, now in the grip of red passion.

An instant, and then horror struck us. And I could feel it surge over the audience. A gasp of horror. The frightened girl in white tried to escape. The musicians wavered and broke. I stared, stricken, with freezing blood. Upon the stage the knife went swiftly up; it came down; then up again. The red Sensua stood gloating. The knife she waved aloft was truly dripping crimson now.

With a choked, gasping scream the white girl of the toilers crumpled and fell.... She lay motionless, at the feet of the crimson murderess.

CHAPTER VIII
"WHY, THIS IS TREASON!"

There was a gasp. The audience sat frozen. On the stage, with no one lifting a hand to stop her, the crimson murderess made a leap and vanished. A moment, and then the spell broke. A girl in the audience screamed. Some one moved to stand up and overturned a seat with a crash.

The amphitheater under the canopy broke into a pandemonium. Screams and shouts, crashing of seats, screaming, frightened people struggling to get out of the darkness. The torches on the stage were dropped and extinguished. The darkness leaped upon us.

Derek and I were gripping Hope. We were struck by a bench flung backward from in front. People were rushing at us. We were swept along in the panic of the crowd.

I heard Derek shout, "We must keep together!"

We fought, but we were swept backward. We found ourselves outside the canopy. Torchlight was here. It glimmered on the pool of water. People were everywhere rushing past us, some one way, some another. Aimless, with the shock of terror upon them. Under the canopy they were still screaming.

I was momentarily separated from Derek and Hope. I very nearly stumbled into the pool. A girl was here, crouched on the stone bank. Her wet crimson veils clung to her white body. Her long, wet hair lay on her. I stumbled against her. She raised her face. Eyes, wide with terror. Mute, painted red lips....

I heard Derek calling again, "Charlie!" I shoved my way back to him. The crowd was thinning out around us. Girls were climbing from the pool, rushing off in terror, to mingle with the milling throng. Among the crowd now, down by the edge of the bay, I saw the sinister figures of men come running. The toilers, miraculously appearing everywhere! I saw, across the pool, a terrified girl crouching. A huge man in a black cloak came leaping. The colored lights in the trees glittered on his upraised knife blade as it descended. The girl fell with a shuddering scream. The murderer turned and whirled away into the crowd.

"Charlie!"

I was back with Derek and Hope. Hope stood trembling, with her hand pressed against her mouth. Derek gripped me.

"That cloak, get it off!" He ripped his crimson cloak from him and tossed it away. He jerked mine off. "Too dangerous! That's the crimson badge of death to-night."

We stood revealed in the clothes of our own world. My business suit, in which that day I had worked in Wall Street. Derek in his swagger uniform. He stood drawn to his full height, a powerful figure. The wires of our mechanism showed at his wrists. They dangled at the back of his neck, mounting to that strangely fashioned electrode clamped to his head. Strange, awe-inspiring figure of a man!

We were momentarily alone under the colored lights of the trees. Hope murmured, "But they will see us—see you...."

Derek's face was grim, but at her words he laughed harshly. "See us! What matter?" He swung on me. "It forces our hand; we've got to come out in the open now! This murder—this king! My God, what a fool to let himself get into such a condition as this! His people—this chaos—what a fool!"

He had drawn his dirk. I realized that I was holding mine. Near us the body of a crimson noble was lying under a tree. A sword was there on the ground. Derek sprang for it, waved it aloft.

I think that no more than a minute or two had passed since the murder. Down by the water the boats were hastily loading and leaving the dock. One of them overturned. There were screams everywhere. Red forms lay inert upon the ground where they had been trampled, or stabbed. But the prowling figures of the toilers now seemed to have vanished.

Derek gestured. "Look at the palace! The garden!"

Beyond the canopy I could see the dim gardens surrounding the palace. I glimpsed the high fence, and the gateway in front. A mob of toilers was there. The guard at the gate had fled. The mob was surging through. Men and women in the vivid garments of the fields, armed with sticks and clubs and stones and the implements of agriculture. They milled at the gate; rushed through; scattered over the garden. Their shouts floated back to us in a blended murmur.

We were standing only a dozen feet from the edge of the pavilion. No one seemed yet to have noticed us. A few straggling lights had come on under the canopy. I could see the dead lying there in the wreckage of overturned seats.

Derek said, "We can't help it—it's done. Look at them! They're attacking the palace!"

This mob springing miraculously into existence! I realized that the toilers had planned that if Sensua were chosen they would attack the festival. The murder of Blanca had come as big a surprise to them as to us....

"Come on! Can you get into the palace, Hope? The king must have gotten back there. Get your wits, girl!" Derek stood gripping her, shaking her.

"Yea, there's an underground passage. He probably went that way."

From the palace gardens the shouts of the mob sounded louder now. And from within the building there was an alarm bell tumultuously clanging.

Hope gasped, "This way."

She led us back into the pavilion. We clambered over its broken seats, past its grewsome huddled figures. Some were still moving.... We went to a small door under the platform. A dim room was here, deserted now. Against the wall was a large wardrobe closet; stage costumes were hanging in it. The closet was fully twenty feet deep. We pushed our way through the hanging garments. Hope fumbled at the blank board wall in the rear. Her groping fingers found a secret panel. A door swung aside and a rush of dank cool air came at us. The dark outlines of a tunnel stretched ahead.

"In, Charlie!"

I crouched and stepped through the door. Hope closed it behind us. The tunnel passage was black, but soon we began to see its vague outlines. Derek, sword in hand, led us. I clutched my dirk. We went perhaps five hundred feet. Down at first, then up again. I figured we were under the palace gardens now, as the tunnel was winding to the left. There were occasional small lights.

Derek whispered to Hope, "The toilers don't know of this?"

"No."

"Where does it bring us out?" I whispered.

"Into the lower floor of the castle. The king must have gone this way. There might be a guard, Derek. What will you do?"

He laughed. "I can handle this mob. Disperse it! You'll see! And handle the king." He laughed again grimly. "There is no Blanca to choose now."

The tunnel went round a sharp angle and began steeply ascending. Derek stopped.

"How much further, Hope?"

"Not far," she whispered.

We crept forward. The tunnel was more like a small corridor now. Beyond Derek's crouching figure, in the dimness I could see a doorway. Derek turned and gestured to us to keep back. A palace guard was standing there. His pike went up.

"Who are you?"

"A friend."

But the man lunged with his pike. Derek leaped aside. His sword flashed; the flat of it struck the fellow in the face. Derek, with incredible swiftness, was upon him. They went down together and before the man could shout, Derek had struck him on the head with the sword hilt. The guard lay motionless. Derek climbed up as we ran forward to join him.

I noticed now, for the first time, that in his left hand Derek held a small metal cylinder. A weapon, strange to me, which he had brought with him. He had not mentioned it. He had produced it, when menaced by this guard. Then he evidently decided not to use it.

He shoved it back in his pocket. He whirled on us, panting. "Hurry! Close that door!"

We closed the door of the tunnel.

"Charlie, help me move him!"

We dragged the prostrate figure of the unconscious guard aside into a shadow of the wall. We were in a lower room of the palace. It seemed momentarily unoccupied. Overhead we could hear the footsteps of running people. A confusion in the palace, and outside in the garden the shouts of the menacing throng of toilers. And above it all, the wild clanging of the alarm bell from the palace tower.

Derek said swiftly, "Get us to the king!"

Hope led us through the castle corridors, and up a flight of steps to the main floor. The rooms here were thronged with terrified people—crimson nobles in their bedraggled finery of the festival. In all the chaos no one seemed to notice us.

We mounted another staircase. We found a vacant room; through its windows we looked a moment, gazing into the garden. It was jammed with a menacing mob, which milled about, leaderless, waving crude weapons, shouting imprecations at the palace. At the foot of the main steps the throng stood packed, but none dared to mount. A group of the palace guards stood on the platform over the moat.

Derek turned away impatiently. "Let's get to the king."

We mounted to the upper story. The castle occupants stared at Derek and me as we passed them. A group of girls at the head of the staircase fled before us.

"The king," Derek demanded, "Which is his apartment? Hurry, Hope, we've no time now!"

We found the frightened king seated on a couch with his counsellors around him. It was a small room in this top story of the castle, with long windows to the floor. I saw that they gave onto a balcony which overlooked the gardens. There were perhaps twenty or thirty people huddled in the room. A confusion existed here as everywhere else—no one knowing what to do in this crisis. And that cursed alarm bell wildly adding to the turmoil. We paused at the doorway.

"Now," whispered Derek. He drew himself to his full height. His eyes were flashing. It was a Derek I had not seen before; he wore an air of mastery. As though he, and not the frightened, trembling monarch on the couch, were master here. And as I stared at him that instant in this primitive chaotic environment, the power of him swept me. A conqueror. The strange electrode clamped to his head gave him an aspect miraculous, awe inspiring.

He strode forward across the apartment. The king was just giving some futile, vague command to be transmitted to his guards down below. A hush fell over the room at our appearance. The king half stood up, then sank back.

"Why—why—who—"

I saw Rohbar here. His long crimson cloak hung from his shoulders, with its hood thrown back. Beneath it, as it parted in front, his leather uniform was visible. A sword was strapped to his waist. He was striding back and forth with folded arms, frowning, but his gaze was very keen. Rohbar was not frightened. He seemed rather to be gauging the situation, pondering how he might turn it to his own ends. He stopped short and swung about to face us. His jaw dropped with surprise, amazement, at our strangeness.

Derek confronted him. His bulk, and huge weight towered even over Derek. The king gasped and sat helplessly staring.

Rohbar spoke first. "Who are you?"

"This mob must be dispersed. Don't stand looking at me like that, man!"

Derek spoke in friendly fashion, but vehemently. "This is no time for explanations."

They were menacing each other. Rohbar's heavy hand fell to his sword, but Derek boldly pushed him away. He faced the king.

"Your Majesty...."

The king stared blankly at him. The title was no doubt strange to this realm, but no stranger than Derek's aspect.

"Your Majesty...."

But the noise from the garden, the confusion which now broke out in the room, and that damnable clattering bell, drowned his words.

The king found his voice. "Be quiet, all of you!" He was on his feet. He demanded of Derek again, "Who are you?"

Derek said swiftly, "I'll show you. I can disperse this mob! Charlie, come."

It seemed as though the gaze of everyone in the room went to me. I drew myself up and flashed defiance back at them. And I followed Derek to one of the balcony windows. He went through it, with me after him. I stood at the threshold, watchful of the room behind us. Rohbar was standing aside, and I saw now the woman Sensua with him. They were whispering, staring at me and Derek.

I had been wondering why, when Sensua must have known that the king would choose her—why she had dared to murder her rival. I thought now—as I saw her with Rohbar—that I could guess the reason. She loved Rohbar, not the king. Rohbar was plotting to put himself on the throne, using Sensua as a lover to that end. He had doubtless persuaded her to this murder, knowing it would arouse the toilers, precipitate this chaos which was what he wanted. Scheming scoundrel! I could not forget the look of desire on his face as he had accosted Hope....

And now Derek appeared, to add an unknown element to Rohbar's plans. There was no way he could guess who or what we were. I saw that he was puzzled, was whispering to Sensua about us, doubtless wondering how to handle us.

I saw too, that there were half a dozen crimson cloaked men here who were not frightened. They had gathered in a group. They stood with hands upon their swords, eyeing me, and watching Rohbar—as though at a sign from him they would rush me.

On the balcony Derek stood with the light from the room upon him. The crowd saw him. The main gateway of the palace was just under his balcony.

The crowd had now started up the steps to where the guards were standing at the top. At the sight of Derek the mob let out a roar, and those on the steps retreated down again.

Derek stood at the balcony rail, silent, with upraised arms, gazing down upon the menacing throng. There was a moment of startled silence as he appeared. Then the shout broke out louder than before. The crowd was milling and pushing, but still leaderless. An aimless activity. Someone threw a stone. It came hurtling up. It missed Derek and struck the castle wall, falling almost at my feet.

Derek did not move. He stood calmly gazing down; stood like an orator waiting for the confusion to die before he would speak.

From the platform, just beneath Derek, the guards were staring wonderingly up, awed, startled. To the right a wing of the building turned an angle. The castle tower was there: it rose perhaps a hundred feet higher than our balcony. On the railed platform-balcony girding its top I saw the figures of other guards standing, gazing down at Derek. The clanging bell up there was suddenly stilled.

I became aware of the king close behind me. His voice rang out: "What are you doing? How dare you?"

Derek whirled, "You fool! To what a pass you have come! Your people in arms against you...."

His violent words brought the king's anger. "How dare you! This is treason!"

I stood alert, with my hand upon my dirk.

There would be conflict here, I felt that we could not hold it off more than a moment longer. My mind leaped to that metal cylinder Derek had concealed. A weapon? Then why did he not have it out now? His eyes were flashing. The aspect of power, of confidence, upon him was unmistakable. It heartened me. I took a step toward him.

He smiled faintly. "Wait, Charlie."

The king gasped again. "How dare you? Why, this is treason! Rohbar, seize him!"

Hope was beside me, her eyes watching the room. Rohbar came striding forward. Derek rasped, "You perhaps have some sense! Lead His Majesty away. Take care of him until this is over."

They stood with crossing glances. And upon Rohbar's face a look, queerly sinister, had come. A smile, sardonic.

He said abruptly to the king, "I think we should let him have his way. What harm?"

He gestured and Sensua came forward. The crimson murderess! Her voluptuous figure was shrouded in a crimson cloak. Her heavy painted lips smiled at the King. Her rounded white arms went over his shoulders.

"Leonto, do as Rohbar says. Let this stranger try. It can do no harm."

The king yielded to her; I watched as she and Rohbar urged him through an archway that gave into the adjoining apartment.

No wonder Rohbar was sardonically smiling! Derek had played into his hand. We did not know it then, but we were soon to find it out.

CHAPTER IX
"ALEXANDRE—"

Derek turned back to the balcony. It had been a brief interlude. The mob in the garden, the soldiers at the top of the stairway, and the other guards high on the bridge of the tower were all standing gazing. Shouts again arose as Derek appeared. Again he raised his arms. This time his voice rang out.

"Silence all of you! I am a friend! Silence!"

At first they did not heed him; then someone shouted:

"Quiet! Listen to him! Let him talk!"

The crowd was bellowing, and then they ceased. The bell was still. In the hush came Derek's voice:

"I am a friend. I come from foreign lands, from distant lands of strange people and strange magic."

For answer the crowd shouted and milled in confusion. A stone came up and then another. Derek stood immovable, like a statue gazing down at them.

"I command you to disperse. You will not? Then look at me! Look at me, all of you. My will is law beyond this king—beyond these palace soldiers— beyond any power you have ever known."

Then I knew a part of Derek's purpose! He had pressed the mechanism at his wrist. He stood imperious with upraised arms. The garden was in a tumult, but in a moment it died. A wave of horror swept the crowd. A freezing, incredulous horror. They stood staring, incredulous, silent, swept with a widening wave of horror.

The figure of Derek on the balcony was fading, turning luminous. A wraith, a ghost of his menacing shape standing there. It faded until it was almost gone, and then, as he reversed the mechanism, it materialized again. A moment passed, then he stood again solid before them.

His voice rang out, "Will you obey me now? I am a friend of the toilers!"

They were prostrate before him. There is no fear more terrible than the fear of the supernatural. In all of history there has been in our world no worship more abject than the worship and fear of a primitive people for its supernatural God. On the platform beneath the balcony, the palace soldiers stared up, horrified. Then they too were prostrate before Derek's threatening gestures and commanding voice.

I stood watching, listening. And suddenly, from the prostrate crowd, a man leaped up. In the silence his amazed voice carried over the garden.

"Alexandre! It is our Prince Alexandre! Our lost prince!"

He stood staring at Derek, his arms gesturing to his comrade around him. He shouted it again:

"Our rightful king, come back to us! Don't you recognize him? *I* saw him go! He went like that—fading into a ghost. Ten years ago, when Leonto killed his father and would have killed him had he not escaped!"

The crowd was standing up now. They recognized Derek! There was no doubt of it. The garden was ringing with the tumultuous shouts,

"Alexandre! Our lost prince has come back to us!"

My head was whirling with it. Derek, prince of this realm? I could see that it was true. Escaped from here as a young lad, when his throne was usurped. Returning now, a man, to claim his own.

And suddenly he turned and flashed me his smile.

The din from the garden drowned his words. The crowd was shouting: "Alexandre! Our lost prince!"

The king's guards on the lower platform stood sullen, confused. I heard footsteps behind me. I whirled around.

From the room, the group of Rohbar's crimson nobles were rushing toward me! Their swords were out. One of them shouted, "Kill them now! We must kill them and have done!"

There were five or six men in the group. They were no more than ten feet away from me. They came leaping.

I stood in the window opening, with only my dirk to oppose them. I shouted, "Derek! Derek!"

I think I took a step backward. I was out on the balcony. It flashed over me—Derek and I were caught out here!

The first of the red cloaked figures came hurtling through the doorway. I leaped to avoid his sword. I saw the others crowding behind him.

Then I felt Derek shove me violently aside. I half fell, but recovered myself at the balcony rail. Five of the crimson nobles were on the balcony. Derek confronted them. His aspect made them pause. They stood, with outstretched swords. The garden was silent; the crowd stared up. And in the silence Derek roared,

"Get back! All of you, go back inside! Back, or I'll kill you!"

In Derek's right hand he held the cylinder outstretched, leveled at the menacing nobles.

"Back, I say!"

But instead they rushed him. There was a flash. From the cylinder it seemed that a ray spat out, a flash of silver light. It caught the three men who were in advance of the others. Their swords dropped with a clatter to the balcony floor. They stood, transfixed.

An instant. Derek's silver ray played upon them. Their red cloaks were painted with its silver sheen.

They were shimmering! I gasped, staring. The other nobles, beyond the ray, had fallen back. And they too stood staring in horror.

Another instant The three figures wavered. I saw the face of one of them, with the shock of incredulous horror still upon it. A face turning luminous! A face, erased, with only the staring eyes to mark where it had been!

There was a moment when the three stricken men stood like shimmering ghosts, with Derek's deadly ray upon them. Then they were gone! It seemed, just as they vanished, that they were falling through the balcony floor....

Derek snapped off his ray. He rasped, "Back into that room, I tell you!"

The remaining nobles fled before him. He turned again to the balcony rail.

"My people—yes, I am Alexandre—I had not thought you would recognize me so soon. But you are right—the time has come for me to claim my inheritance. And I will rule you justly."

His cylinder was still in his hand; he swept a watchful glance behind him. I thought of Rohbar. He was in the next room, with the king. Had they seen this attack upon Derek? They must have heard the crowd shouting, "Alexandre!" It seemed strange they did not appear.

I recall now, as I look back to this moment on the balcony, that I suddenly thought of Hope. She had been beside me just before the nobles attacked. I did not see her now. I was startled, but thought of her was driven from my mind. From within the palace a scream sounded. A girl screaming.

But it was not Hope's voice. A girl, screaming, and then shouting:

"The king is dead!"

Derek came rushing at me. "Charlie, that—"

We heard it again. "The king is dead!"

We hurried into the adjoining room. There was no one to stop us—no one up here now who dared oppose Derek. The terrified nobles in the room fell cringing before him.

"Alexandre—spare us! We are loyal to you!"

He strode past them. In the adjacent apartment we found the king lying upon the floor. A wound in his throat welled crimson. He had evidently been lying here alone, and had just now been found by a girl who had entered. He was not quite dead. Derek bent over him. He opened his eyes.

He gasped faintly: "Rohbar—killed me. Rohbar and that—accursed crimson Sensua...."

His voice trailed away. The light went out of his staring eyes. Derek laid him gently back on the floor.

And as though already the news of his death had miraculously spread, the bell in the castle tower began tolling. Not clanging now. Tolling, with slow, solemn accent. The crowd evidently recognized it. We could hear the shouts: "Death! Death has come!"

Derek's eyes ware blazing as he stood up. "The end, Charlie! I would not have planned this, and yet...."

He did not finish. He whirled, rushed back to the other room and to the balcony. The scene was again in confusion the crowd milling, voices shouting:

"The king is dead!"

At the edge of the garden a woman's shrill, hysterical laughter rose over the din.

Derek called, "Yes, the king is dead!" He paused. Then he added, "If you want me—if I have your loyalty—I will claim my throne."

A tumult interrupted him. "Alexandre! King Alexandre!"

He spread his arms, but he could not silence them.

"The king is dead. Long live King Alexandre!"

A wave of it swept over the garden, engulfing the castle. At the main entrance Leonto's soldiers stood sullen, listening to it.

Derek stood triumphant. His hands were outstretched, palms down. But up on the circular bridge at the top of the tower there was a sudden commotion. The soldiers up there had vanished, moved back within the tower to make room for other figures. I stared amazed, transfixed. A huge man in leather garments was there, with a sword stuck in his wide belt. A man with a bullet head, a heavy face, gazing down....

Rohbar!

And held in front of him the slender figure of a girl. Hope! He clutched her, his thick arm encircling her breast. With sinking heart I realized what had happened. Hope had moved away from me. Every one in the room had been intent upon Derek. Rohbar had come quietly in, after murdering the king, had seized Hope, stifled her outcry, and had taken her up into the tower.

And I had promised Derek that I would shield this girl from harm! The horror of it—the self-condemnation of it—swept me, froze me to numbness. I could not think; I could only stand and stare. Rohbar held Hope like a shield before him. The low railing hardly reached her knees. A sheer drop to the garden beneath. He held her tightly, and in his free hand I saw his dirk come up menacingly against her white throat. His voice called:

"Silent, down there! Alexandre, you traitor! Silence!"

Derek stared up. The triumph faded from him. He stared, stricken. The crowd stared. The soldiers on the lower platform ceased their shouting and gazed up at these new actors, come so unexpectedly upon the stage. Again Rohbar called, to the guards this time:

"I represent your King Leonto. This Alexandre is a traitor to us all. And he cannot harm me! I defy him. Look at him! I defy him to use his evil weapon upon me!"

Derek was silent. A single adverse move and Rohbar's knife would stab into Hope's throat. Derek's ray was powerless. A flash from it would have killed Hope, not Rohbar.

The king's soldiers saw Derek's indecision. One of them shouted, "He cannot harm us! Look, he is frightened!"

The crowd recognized Hope. They began calling her name. And calling, "Master Rohbar, do not harm our Hope!"

"I will not harm her! Not if you do what I tell you! Leave the garden—go quietly! I will deal with this traitor!"

He added to the guards, "Go up and seize him! He cannot hurt you! Traitor! Seize him! If he does not yield—if any of this crowd attacks you— then I will kill Hope."

Derek stood clinging to the balcony rail. With Rohbar's watchful gaze upon him he did not dare turn or move. I was standing back from the balcony, behind Derek and partly in the room. No one thought of me. No one from outside could see me. And I, who had played no part in this, save that one I had neglected, suddenly saw my role. My cue was sounding. My role to play, here upon this tumultuous stage.

I turned back into the dim room. A few frightened men and girls were here. They were all crowding forward, gazing through the windows at the scene outside. No one noticed me, but I saw, with sudden realization, my role to play.

I darted across the room, out into the dim, deserted corridor of the castle.

CHAPTER X
MY ROLE TO PLAY

I slipped like a shadow through the almost empty corridors. Down on the lower floor I found that many of the soldiers were on the inside, standing about the corridors in groups, waiting for word from their comrades on the platform to indicate what action they should take. My time was short; I knew that within a few minutes they would be rushing up to overpower Derek.

I stood unseen against the wall near the main entrance. I could not get outside. There were too many soldiers there.

I tried to keep my sense of direction. The wing upon which the tower stood was about two hundred feet from me here. If I could not get outside I would have to try the inside, along this corridor. I prayed that I might not make an error. I tried to gauge exactly where the tower would be.

The hallway was almost dark and in this wing there chanced to be no one at the moment. I came to the angle and turned it to the left. I was unarmed save my dirk. I drew it. But I encountered no one. I passed the doors of many empty rooms. The windows were all barred on this lower floor. I could hear the shouts of the crowd outside.

I came at last to the end of the wing. A staircase here led upward. I guessed that I was directly under the tower now, and that this staircase undoubtedly led upward into it. I mounted a few steps to verify what I was sure would be the condition. It was as I thought. Rohbar had won over the soldiers who were here. He had sent them down from the tower bridge. They were guarding this staircase.

I crept up another few steps, very cautiously. I could hear their voices on the stairs. A light was up there. I could see the legs of some of them as they crowded the stairs. I softly retreated.

There was no way of getting up into the tower here. Alone and armed only with my dirk, I could not mount these stairs and assail a dozen armed men standing above me; especially when, if I raised an alarm, Rohbar overhead might be startled into killing Hope.

I stood another moment, thinking, planning my actions. I was trembling. Everything depended upon me now. I must get up into the tower. And, above everything, haste was necessary.

I retreated back to the lower floor. I was still some twenty feet above the ground, I judged. That was too far. A dozen paces along the hall I saw a stairway leading downward into the ground level cellar of the castle. I marked in my mind exactly in which direction I turned, and how far. I went down the stairs.

There was an empty lower room. It was pitch black. I lay down on its earthen floor. Above me, a few paces off to one side I could visualize the tower. A hundred and fifty feet above me, at least, up to that bridge balcony, where Rohbar stood with Hope. I kept my mind on it and prayed that I might not be making an error, a miscalculation.

I prayed, too, that luck would be with me. A desperate chance, yet I thought I knew what was here, or about here, in New York City. I lay on my side, alone in the blackness, and pressed the switch at my wrist....

The familiar sensation of the transition began. The darkness grew luminous. Around me shadows were taking form. My body was humming, thrilling with the vibrations within it. I could feel the ground under me seeming to melt. My head was reeling. Nausea swept me, but with it all I tried to keep my wits. I must watch this new Space into which I was going. Space? I prayed that here on this spot in New York City there would be empty space! If not, at the first warning, I was prepared to stop my mechanism.

The shadows grew around me. There was a moment or two when I felt as though I were floating. Weightless. The sense of my body hovering in a void, intangible, imponderable, with only my struggling mentality holding it together....

And then I felt myself materializing. Around me walls were taking form. I floated down a foot or two and came to rest upon a new floor. My hand brushed it. My physical senses were returning. I could feel a floor of concrete. A vague, shimmering light was near me. It seemed to outline the rectangle of a window. All around was darkness. Empty darkness. Soundless, with only the throbbing hum of the mechanism....

I was indoors, in a room. I felt suddenly almost normal, except for the whirring vibration. I flung the switch again. There was a shock. A whirling of my senses. Then I sat up; my head steadied. The nausea passed.

I was back in my own world, in New York City. This was night: I tried to calculate the time. Derek and I had departed about midnight. This would be, then some time before dawn. I was in a cellar room, lying on its cement floor. There was a window, with a faint light outside it. A window up near the ceiling. A straggling illumination showed me a bin, a few barrels, a door leading into another room which looked as though it might be a machine shop.

I sat up, calculating. I was a thousand feet perhaps from the Battery wall, two hundred feet from the Hudson River. This was an office building, and I was in one of its cellar rooms, at the ground level.

Near dawn? I tried to calculate what might be overhead. A deserted office building. Too early yet for the scrub-women. The elevator would not be running. I laughed to myself. Of what use to me an elevator, if it had been running? How could I, a midnight prowler, appear from the cellar of this building, and demand to be taken upstairs! There would be no elevator, but there would be watchmen. I would avoid them.

I found a door. My heart leaped with a sudden fear that it would be locked, but it was not. I went through it into a passage and found the staircase. I made two turns. I tried again to keep my mind on this Space here. I stood, carefully thinking. I had it clear. I had made no move without careful thought. The tower with Rohbar was still to my left, and about directly above me.

I went up the short stone staircase, opened another door carefully. I was in the dim lower hall of the office building. I found myself beside the deserted elevator shaft. A light was burning on the night attendant's table in an alcove, on the other side of the shaft. He sat there with his back to me. I closed the door soundlessly.

The stairway upward beside the elevator was here. I watched my chance. I darted around the angle and went up. I met no one. The concrete staircase had a light at each floor. Four floors up. No, not enough! I opened the fourth floor door. The marble hall of the office building was empty and silent. Rows of locked office doors with their gold-leaf names and numbers. A single dim light to illumine the silent emptiness....

I retreated into the staircase shaft and mounted higher. My dirk was in my hand. Charlie Wilson, the Wall Street brokerage clerk, prowling here! And upon what a strange adventure!

I came to what I thought was the proper floor. In the hall I selected a room. The door was securely locked. I had no way of breaking the lock, but the panel was of opaque glass. I would have to chance the noise. I rushed the length of the hall, to where a red fire-ax hung in a bracket. I came back with it. I smashed the glass panel of the door.

Would a watchman hear me? I did not wait to find out. With the ax I scraped away the splinters of glass. I climbed through the opening. My hand was cut, but I did not heed it.

I was in a dim, silent office, with rugs on the floor, desks standing about, filing cases, a water-cooler, and a safe in the corner. I rushed to one of the windows. It looked over Battery Park and the upper bay. The stars were shining, but to the east over Brooklyn I could see them paling with the coming dawn. I gazed down to try and calculate my height. Yes, this would be about right. And my position. I could see the outline of the shore, the trees of Battery Park, the busy harbor, even at this hour before dawn, thronged with the moving lights of its boats.

I saw all this with my eyes, but with my mind I saw the wrecked, deserted pavilion, and the gardens of Leonto's castle. The threatening mob would be below me. The palace entrance would be here to my left, down in the street where those taxis were parked. There was a commotion down there by the office building entrance. I know now what caused it, but at the time I did not notice. The wing of the castle was under me. This would be the tower. Its upper room, or the balcony, just about where I was standing. I prayed that it might be so. I seemed with my mind to see it all.

I lay down on the floor by the window. Out in the office building hallway I heard heavy footsteps come running. One of the night watchmen had evidently heard the glass crashed.

I laughed. I pressed the switch at my wrist....

CHAPTER XI
THE FIGHT ON THE TOWER BALCONY

The sensations swept me again. The room faded. Whether the watchmen came in to see a ghost of me lying there on the floor I did not know, nor did I care. I whirled into the shadows. And came in a moment out of the black silence. The office room was gone. I seemed to have fallen or floated down—how far I do not know. A triumph swept me. I was lying on another floor. I could see a doorway materializing. I was not upon the balcony as I had calculated, but within the tower room. New walls sprang around me.

I did not heed it, this time, the sensation, of the transition. I was too alert to what new situation might come upon me. The tower room. I could see it. I could see its oval windows close at hand. The doorway to its balcony. Sounds flooded me, mingled with the humming within me. Familiar sounds. The crowd shouting. And a single voice—the voice of Rohbar. Vague and blurred, but as I materialized it became clearer.

I was suddenly aware that there was a man beside me. One of the palace soldiers. He saw me materialize. He leaped backward in horror. I flung my switch. I was on my feet, swaying, and then I leaped upon him. My dirk plunged downward into his chest.

The thing made me shudder. I reeled with the sickness of it, but as he fell I clung to the dirk and ripped it out of him. It was dripping with his blood.

I stood trembling. The small tower room had no other occupants. I turned toward the door. I could see a patch of stars, paling with the coming dawn. I crouched in the small doorway which gave onto the balcony, staring, swiftly calculating. The scene had scarcely changed. But, some of the soldiers had left the entrance platform, gone, no doubt, into the castle on their way upstairs to seize Derek.

On this upper balcony, no more than ten feet before me, Rohbar still stood gripping Hope. She was in front of him. His back was to me. A sudden jump, and I could plunge my dagger into his back.

Rohbar was shouting, "King Leonto is dead. If you should want me to succeed him, I will take this girl Hope for my queen. You all love her...."

I was tense to spring. Then out in the balcony, to one side, I saw Sensua crouching. Her crimson robe fell away to bare her white limbs. Her hand fumbled in her robe. She had been Rohbar's dupe, and now she knew it. Her knife was in her hand. Frenzied with jealousy and rage she sprang upon Rohbar's back, trying to stab at Hope.

Perhaps he sensed her coming, heard her; or perhaps she was unskilful. Her knife only grazed Hope's shoulder. He released Hope. He roared. He turned and gripped his murderous assailant. A second or two while I stood watching. He caught Sensua's wrist, twisted the knife from it and plunged the knife into her breast. She sank with a scream at his feet, and as he straightened he saw me.

But I had leaped. I was upon him. His own knife had remained in Sensua's breast. As I raised mine in my leap, he caught at my wrist; twisted it, but I flung the knife away before he could get it. The knife fell over the balcony rail. The weight of my hurtling body flung him backward, but the rail caught him. His arms went around me. Powerful arms, crushing me. I gripped at his throat.

There was an instant when I thought that we would both topple over the railing. I felt Hope beside us. I heard her scream. We did not go over the rail, for Rohbar lurched and flung us back. We dropped to the balcony floor, rolling, locked together. He was far stronger and heavier than I. He came uppermost. He lunged and broke my hold upon his throat, but I was agile: I squirmed from under him. I almost regained my feet. He got up on one knee. He was trying to draw his sword. Then again I bore into him, kicking and tearing. He roared like a bull. And ignoring my plucking fingers, my flailing fists, he lunged to his feet with me gripping again at his throat.

His huge height swung me off the ground. I was aware that he had drawn his sword, but I was too close for him to use it. He swayed drunkenly with my weight; he was confused. I felt the rail behind us. We lunged again into it. Again I heard Hope scream in terror, and saw her leap at us. Rohbar stooped, trying to clutch the low rail. His bending down brought my feet to the balcony floor. With a last despairing effort I shoved him backward. And as he toppled at the rail, I fought to break his hold upon me. I felt us going and then I felt Hope reach me. Her arms flung about my waist. Her hold tore me loose. Rohbar's huge body fell away....

For an instant Rohbar seemed balanced upon the rail; then he went over. He gave a last long, agonized scream as he fell. I did not look down. I crouched by the rail. The crowd in the garden; Derek standing on the other balcony; the soldiers who now had appeared behind him—all were silent, and in the silence I heard the horrible thud of Rohbar's body as it struck....

I clung to Hope for an instant, and she shuddered against me. The scene broke again into chaos. I cast Hope away and leaped up. I stood at the balcony rail. My arms went up and gestured to Derek. Amazement was on his face, but he answered my gesture. Behind him the soldiers who had come to seize him were standing in a group, stricken at this new tragedy.

Derek swung on them. He was not powerless now! "Away with you!"

His cylinder menaced them, and they fell back in terror before him.

He darted past them and disappeared into the castle.

I felt Hope plucking at me. "I want to talk to the people."

She stood beside me, leaning over the rail. Gentle little figure. Familiar figure to them all. Their beloved Hope. Her voice rang out clearly through the hush.

"My people, we all want our beloved Alexandre—he has come back to us. He is our rightful king."

"King Alexandre! Long live King Alexandre!"

Derek in a moment appeared behind us. "My God, Charlie, I can't understand—"

I told him how I had done it. He gripped me. "I'll never be able to repay you for this!"

I pushed him forward and he joined Hope at the rail. Held her, and her arms went around his neck as she returned his kisses. The crowd gaped, then cheered.

I shouted, "Hope will be your queen—The reign of the crimson nobles is at an end!"

The wild cheering of the people, in which now the castle guards were joining, surged up to mingle with my words.

CHAPTER XII
ONE TUMULTUOUS NIGHT

I come now with very little more to record.

I returned to my own world. And Derek stayed in his. Each to his own; one may rail at this allotted portion—but he does not lightly give it up.

The scientists who have examined the mechanism with which I returned very naturally are skeptical of me. Derek feared a further communication between his world, and mine. He smiled his quiet smile.

"Your modern world is very aggressive, Charlie. I would not want to chance having my mechanism duplicated—a conquering army coming in here."

And so he adjusted the apparatus to carry me back and then go dead. I have wires and electrodes to show in support of my narrative. But since they will not operate I cannot blame my hearers for smiling in derision.

Yet there is some contributing evidence. Derek Mason has vanished. A watchman in an office building near Battery Park reports that at dawn of that June morning he heard splintering glass. He found the office door with its broken panel, and the ax lying on the hall floor. He even thinks he saw a ghost stretched out by the window. But he is laughed at for saying it.

And there is still another circumstance. If you will trouble to examine the newspaper files of that time, you will find an occurrence headed "Inexplicable Tragedy at Battery Park." You will read that near dawn that morning, the bodies of three men in crimson cloaks came hurtling down through the air and fell in the street near where several taxis were parked. Strange, unidentified men. Of extraordinary aspect. The flesh burned, perhaps. All three were dead; the bodies were mangled by falling some considerable height.

An inexplicable tragedy. Why should anyone believe that they were the three crimson nobles whom Derek attacked with his strange ray?

I am only Charles Wilson, clerk in a Wall Street brokerage office. If you met me, you would find me a very average, prosaic sort of fellow. You would never think that deeds of daring were in my line at all. Yet I have lived this one strange tumultuous night, and I shall always cherish the memory.

.

The Stolen Mind

By M. L. Staley

What would you do, if, like Quest, you were tricked, and your very Mind and Will stolen from your body?

"What caused you to answer our advertisement?" Owen Quest felt the steel of the quick gray eyes that jabbed like gimlets across the office table.

"Why does any man apply for a job?" he bristled.

Keane Clason gave an impatient smile.

"Come!" he said. "I'm not trying to snare you. But there were unusual features to my ad, and they were put there to attract an unusual type of man. To judge your qualifications, I must know just why this proposition appeals to you."

"I can tell you that," nodded Quest, "but there's nothing unusual about it. In the first place, I knew that the Clason Research Corporation is the leading concern of its kind in the country. In the second place, this seemed to offer a way to obtain a substantial sum of money quickly."

"Good," said Clason. "And you feel that you have all the necessary qualifications?"

"Decidedly. I am 24 years old, athletic, and of an earnest and determined nature. Moreover, I have no family ties, and I'm willing to run any reasonable risk in order to improve the condition of my fellow men."

Clason smiled his approval.

"You say you need money. How much immediately?"

Quest was unprepared for the question.

"A thousand dollars," he ventured.

Without hesitation Clason counted out ten one-hundred-dollar notes from his wallet and laid them on the table.

"There's your advance fee. You're ready to go to work immediately, I hope?"

"Certainly," stammered Quest.

Stunned by the swiftness of the transaction, he sat staring at the money that lay untouched before him.

To accept it would be like signing an unread contract. But he had asked for it; to refuse it was impossible. Even to delay about picking it up might arouse Clason's suspicion. Already the latter had turned away and was opening the door of a steel cabinet. Quest had one second in which to reach a decision.... He crammed the currency into his pocket.

With delicate care Clason set two objects on the table. One looked to Quest like a miniature broadcasting tower or a mooring mast for lighter than air craft. The other was a circular vat of some black material, probably carbon. Within it a series of concentric tissues were suspended from metal rings, and in a trough outside ranged four stoppered flasks containing liquids of as many different colors.

"Look at these models carefully," said Clason. "They represent two of the most remarkable discoveries of all time. The one on your left is the most *de*structive weapon known to man. The other I consider the most *con*structive discovery in the history of science. It may even lead to an understanding of the nature of life, and of the future of the spirit after death.

"Both of these were developed by my brother Philip and me together— but we have disagreed about the use to which they shall be put.

"Philip"—the inventor dropped his voice to a whisper—"wants to sell the secret of the Death Projector—the tower, there—as an instrument of war. If I should permit him to do that, it might lead to the destruction of whole nations!"

"How?" demanded Quest "I've heard of a device called the Death Ray. Is this it?"

"No, no," said Clason contemptuously. "Even in a perfected state the Ray would be a child's toy compared to the Projector. This is based on our discovery that invisible light rays of a certain wave-length, if highly concentrated, destroy life—and our additional discovery that if these are synchronized with short radio waves the effect is absolutely devastating.

"We obtain the desired concentration of invisible light by using a tellurium current-filter under the influence of alternate flashes of red and blue light. The projector can literally blanket vast areas with death, up to a top range of at least five hundred miles.

"Just picture to yourself what this means! In a space of ten minutes two men can lay down a circle of destruction a thousand miles in diameter; or they can cut a swath five hundred miles long in any desired direction."

"Have you ever proved it?" demanded Quest skeptically.

"Yes, young man, we have," snapped Clason. "Right here in the laboratory—but on a minute scale, of course. However, there's no time to demonstrate now. The point is that my brother is determined to sell if he can obtain his price for the invention. He argues that instead of bringing disaster upon the world, this machine will forever discourage war by making it too terrible for any civilized nation to consider. In spite of my opposition he has opened negotiations with an ambitious Balkan power. He may actually close the sale at any moment!

"However," Clason drew a deep breath "you see this other device? Simple as it appears, it is the key to the whole situation. We can use it—you and I—to overcome Philip's will and prevent this unthinkable transaction. The two of us can do it. Alone I would be virtually helpless."

"Why not have the Projector confiscated or destroyed by our own Government?" suggested Quest. "That seems to me the only safe and sure way out of the difficulty."

"You simply do not understand," frowned Clason impatiently. "Philip is selling the plans and descriptions of the machine, not the machine itself. Even if this model and the larger test machine that we have built were destroyed—even if I were willing to have Philip sent to Leavenworth for life—he could still sell the Projector.

"But this other invention, our Osmotic Liberator, makes it possible for me to gain control of Philip and actually *change his mind*, through the medium of an agent. I have hired you to act as my Agent, Quest, because I can see that you are a young man of unusual character and vitality. And by way of reward I can promise you both money and a brilliant future."

The inventor poised in a tense attitude on the edge of his chair as though his body were charged with electricity. His eyes seemed to dart out emanations that set Quest's blood to tingling. Then for a moment the latter lost consciousness of his physical self. It was as though he had opened a door and found himself suddenly on the brink of a new and totally strange world. He dispelled this fancy by a quick effort of the will, for he knew that he had a delicate problem on his hands and that it must be solved within a very few minutes. However he proceeded, he must act without disloyalty to his Government, and at the same time without injustice to Keane Clason.

"Tell me," he said in a husky voice, "how do you intend to use me? I do not believe in Spiritualism. I would be a poor medium."

Clason gave a short laugh.

"You are not to be a medium in that sense at all. Spiritualism as practiced is just a blind sort of groping and hoping. Osmotic Liberation, on the other hand, is an exact and opposite physico-chemical science. Here—I will show you."

Into the outer cell of the Liberator he emptied the purple vial, and so on to the innermost, which he filled with a golden-green liquid like old Chartreuse.

"The separating membranes, you understand, are permeable by these complicated solutions. Each liquid has a different osmotic pressure and therefore should, under normal conditions, interchange with the others through the membranes until all pressures are equalized. I prevent such interchange, however, by maintaining an anti-electrolysis which retards ionization and thus builds up what might be called osmotic potential.

"Now if an Agent—yourself for instance—submerges himself in the central cell, at the same time maintaining a physical contact with his Control at the surface of the liquid, and if then the osmotic potential is suddenly released by throwing the electrolytic switch, the host of ions thus turned loose in the outer compartments make one grand rush for the center solution, which contains the cathode.

"Under these conditions your body becomes a sort of sixth cell, and your skin another membrane in the series. Properly speaking, however, you are not a part of the electrolytic circuit but are merely present in the action. Your body acts as a catalyser, hastening the chemical action without itself being affected in any way. Physically you undergo no change whatever; but in some strange way which is, like life, beyond analysis, your mind flows out into the solution, while your unaltered body remains at the bottom of the tank in a state of suspended animation.

"If no Control is present, all that is needed to return your mind into your body is a throw of the electrolytic switch back to negative, whereupon you emerge from the tank exactly as you entered it. But with your Control present and in contact with your submerged body, your mind, instead of remaining suspended in the solution, flows instantly into his body and resides there subject to his will.

"This can not be done, however, unless the wills of Control and Agent have first been brought into accord. To accomplish that, we clasp hands"—

Quest grasped Clason's extended hand—"and look steadily into each other's eyes.

"Now, it is well known that the vibrations of an individual's will are as distinctive as the sworls of his finger-prints. What is not so well known is that the frequency of vibration in one person can be brought into accord with that in another.

"You consciously retract your will by concentrating your mind upon the thing which you know I wish to accomplish. Gradually while we continue in this position your vibrations speed up or slow down until they acquire exactly the same frequency as my own. We are then in accord, and when your mind is liberated in the tank it is in a state which admits absorption into my body. And it is subject to my will because you have purposely attuned it to my peculiar frequency. Immediately after the transfer there will be a brief conflict, due to the instinctive desire of your will to obtain the ascendancy. But of course mine will gain the upper hand at once, since both wills will be in my frequency."

Quest felt, rather than saw, a wall of alarm closing in on him. He tried to avert his eyes, to withdraw his hand from Clason's grasp. With a nostalgic pang in the pit of his stomach he suddenly realized that he could not do so. He had gone too far—farther than any man in his position had a right to go. Having deliberately weakened his will, it seemed now to have deserted him entirely. A prickling sensation coursed up his spine, his extended arm went numb, his hand trembled violently.

"Splendid!" said Clason, suddenly releasing both eye and hand. "Just as I foresaw, you will be able to attune yourself to my vibration-frequency with hardly an effort. Now please remain seated; I'll be back in a moment."

For a second after the door closed, Quest remained slumped in his chair. Then he was on his feet, shaking himself like a wet dog to free himself from the spell under which he had fallen. Something about Clason attracted and at the same time repelled him, fraying his nerves like an irritant drug and confusing his mind at the moment when he needed the full alertness of every faculty.

Invisible light—disembodied minds—will vibrations! Nothing there to get hold of. Were these things real or imaginary? Was Keane Clason a great inventor, or a madman? Would Philip prove to be a real or an imaginary scoundrel? Should he summon help, or go on alone?

Professional pride said: wait, don't be an alarmist! With his knuckles Quest tapped the table, half expecting it to melt under his fingers. The feeling and sound of the contact gave him a peculiar start. On the farther end of the table stood a letter-box—an invitation. From his pocket Quest snatched a slip of paper, and wrote:

6 stroke 4—9:45A—Hired. If no report in 48 hours, clamp down hard.

To address a stamped envelope and slip it in with the outgoing mail was the work of seconds. But he was none too quick. He had just dropped back into a lounging attitude when the door burst open and Clason flew into the room?

"We must act instantly," hissed the inventor. "Philip plans to close the transaction within a day."

In spite of himself, Quest jumped upright in his chair. Clason tapped him on the shoulder reassuringly.

"It's all right," he smiled, "I'm ready for him. We'll make our move this afternoon and beat him by eighteen hours.

"Let's see." He paused. "Oh! yes. I was about to explain to you that as soon as the will of the Agent enters the body of his Control, the latter can again transfer it into the body of still another person.

"Now you understand why I advertised for a man of exceptional character? As my Agent, I want you to enter the body of Philip, and your will must be strong enough to conquer his in the battle for mastery which will begin the instant you intrude into his body. You will still be under my control, but your will must be strong enough on its own merits to overcome his. I can direct you, but your strength must be your own. That's clear, isn't it?"

"I think so," said Quest slowly. "But what becomes of me after you have frustrated Philip's plot?"

"That's the easy part of the process," smiled Clason; "but naturally you feel some anxiety about it. I simply withdraw your will from Philip, return it to your own body, and pay you a reward of ten thousand dollars."

"You're sure you can?"

"Perfectly. I have merely to touch Philip's hand to recapture your will. Then I immerse myself in the tank with the switch at plus. The osmotic action will extract both wills momentarily from my body. But the presence of two bodies and two wills in the solution together forces a balance, and each will seeks out and enters its own body. Then you and I climb out of the tank exactly as we are this minute."

"If it weren't for my belief that anything is possible," Quest shook his head, "I'd say that your claims for this invention were ridiculous."

"And you couldn't be blamed," admitted Clason readily. "This toy of a model is hardly convincing. But come along with me and I'll show you how the Liberator looks in actual operation."

The office rug concealed a trap door which gave upon a spiral stair. Below, Clason unlocked another door and led the way through a narrow and tremendously long passage lighted at intervals by small electric bulbs. Presently another door yielded to the inventor's deft touch and closed behind them with a portentous chug. Here the darkness was so utter and intense that Quest imagined he could feel the weight of it on his shoulders. From the slope of the passageway and the muffled beat of machinery that had come to his ears on the way along, he guessed that he was below ground in some chamber at the rear of the factory.

He gave a low exclamation as Clason switched on the toplight. No wonder the darkness had seemed of almost supernatural quality! Even the hard white glare of the daylight arc was grisly. Its rays rebounded from the liquids of the great circular tank in a blinding dazzle of color, while the dull black walls and ceiling were so perfectly absorptive that beyond arm's length they became to all effects invisible. Even the ledge on which he stood—the shoulder of the vat—gave Quest the feeling that to move would be to step off into a bottomless pit.

But Clason took his attention at once, pointing here and there in his quick, nervous way to indicate how faithfully the Liberator had been reproduced from the model. In all respects the arrangements were the same, with the addition that here a long plank like a spring-board extended out from a wall-mount as far as the central compartment of the tank, and that from its end a narrow ladder hung down to the surface of the Chartreuse liquid. A double-throw switch fixed to the wall above the base of the plank was evidently the source of electrolytic control.

"When you throw the switch to plus," said Clason, pointing to the chalk-marked sign above, "you produce the violent electrolytic action needed to bring about a liberation. All the rest of the time it should be closed at minus, in order to maintain the anti-action which I explained to you.

"Now let's rehearse, so that when the time for the real performance arrives we can be sure of running it off without a hitch."

"All right, sir," nodded Quest, so dazed by the glittering light that he was hardly conscious of what he said.

"First," said Clason, running lightly up the steps to the plank, "you walk out to the end, like this, and start down the ladder. Then you lower yourself into the tank. The liquid is at body temperature; it's neither strongly acid nor caustic; it will cause you no injury or discomfort whatever.

"Meanwhile I keep in contact with your hand until the instant that you become submerged. Now your mind is in me, see?—ready for transfer into Philip, where it will act as my Agent. That's how simple it is! Come on up and we'll go through the motions."

Quest experienced a shiver as he mounted the bridge. Annoyed with himself, he shrugged the feeling off. There was no risk here. Moreover, it was a part of his daily work to take chances; he had done so a hundred times without hesitation. Now he moved all the more quickly, as if to belie the squeamishness that possessed him in spite of himself.

Swinging past Clason on the plank, he lowered himself without a pause to the bottom rung of the ladder, while the inventor, hanging head down, maintained contact with him.

"No need to stay here," he said in sudden irritation. "I understand perfectly what I am to do."

"I'm testing my own acrobatic ability," grunted Clason amiably. "Just a minute now."

He wriggled as if trying to adjust himself to a better balance, but in reality to mask the motion of his free hand with which he reached up and pressed a button in the side of the plank. Instantly the structure, pivoting downward on its wall-socket, plunged Quest to his waist in the osmotic solution.

"For God's sake get out of the way!" he shouted, trying to wrench his hand out of Clason's sinewy grip. "Let go, I tell you!"

But Clason clung like a leech, his teeth gritted under the strain. Again the plank lurched downward, and with a violent splash Quest vanished below the surface.

Quick as a cat, Clason scrambled up the ladder and back to the base of the plank, where he erased and interchanged the chalk-marked signs with which he had misled Quest. Then with a sinister twist of a smile he threw the switch to minus, and turned to watch as the plank slowly righted itself and the vacant ladder came clear of the liquid.

For some time he stood staring at the gleaming colored rings of his dissociation-vat like some witch over her cauldron, his lips working, his hands clasping and unclasping like the tentacles of some sub-sea monster.

Then, as if the spell had suddenly broken, he turned on his heel and switched off the light. As he hastened down the passageway toward his office, the airlock sucked the door against its jamb with an ominous whistle.

In a twinkling, as Quest's shackled spirit writhed in its new housing, he knew that he was in bondage to a scoundrel. Formless and voiceless, he still fought madly for the freedom which the instinct of ten thousand generations made necessary to him.

At the same time he was furious at himself for having been tricked like an innocent schoolboy. The plank socket, the button which had tripped the supporting spring, the fake rehearsal, the tuning of his will to that of Clason—step by step the whole cunning scheme unfolded itself to him now.

But what could be the purpose behind this villainy? Only one answer seemed possible. Keane must be the one bent on selling the Death Projector, Philip the one who wished to frustrate the fiendish transaction! And Quest of the Secret Service—he was to be the tool to force the sale.

With the soundless scream of rage Quest's will hurled itself against Keane's. The two met like infuriated bulls, and for an instant too brief to be pictured as a lapse of time they poised immovable. But two wills can not exist on equal terms in a single body, and in this case the vibration of both was that of Clason. Quest had challenged the Master Will. He could do no more. It hurled him back, crushed him like foam, compressed him to the proportions of an atom in the background of his consciousness. So brief and unequal was the conflict that in the next breath Clason had all but forgotten the presence of the stolen will within him. When he was ready to use his Agent, that would be time enough to summon him!

Despite this suppression, Quest began to see dimly through strange eyes, and to hear vaguely with ears that were not his own. Feelers, tentacles, some intangible kind of conduits carried thought impulses to him from the Master Will. He received these impressions vividly, but those which he gave off in return were so weak, due to the subjection of his will, that Clason was entirely unconscious of any response. Quest was not enough of a scientist to be astonished at the ability of a disembodied mind to experience sense impressions in the body of another. He was only glad that the darkness and silence were growing less. Very, very slowly he was awakening to a new kind of consciousness—the consciousness of another person's Self. He hated and loathed that Self, yet it was better than the awful blankness that had gone before.

Suddenly, as light grew brighter and sound more clear and definite, a new element entered—the element of hope. At first it was feeble: its only suggestion was that sometime, somehow, he might escape this prison. But it was like water to a parched plant. It caused his will to expand, to extend its feelers, to press up a little more bravely against the crushing pile of the Master Will.

Now another surprise sprang upon him. He was moving! That is, Clason's body was moving in some kind of a conveyance, which was threading its way through crowded streets. Stores, buildings, buses, people—Quest remembered them all distantly as things he had known thousands of years ago. The driver turned his head, and his profile seemed vaguely familiar.

Now a rush of foreign thoughts drowned out his own. They were a sort of overflow from the mind of Clason. They thronged along the conduits that bound the two wills together, but only Quest was conscious of the movement.

Keane's mind was on his brother Philip: that much was particularly clear. And there was something about a telephone call. Yes, Keane had telephoned to the police, disguising his voice, refusing to divulge his name. He had said that a man by the name of Philip Clason was in trouble and had told them where to find him. Then the police had telephoned the factory, and Keane had pretended astonishment and alarm at the news. That's why he was here now—he was on the way to confer with the police. And he was chuckling—chuckling because he had fooled Quest and the police, and because now the hundred million dollars was almost in his grasp.

Cutting in close, the car turned a corner and drew up before one of a row of loft buildings in a section of the city which Quest failed to recognize. As Clason stepped to the sidewalk, Quest was more painfully aware than ever of his powerlessness to influence by so much as the twitch of a muscle the behavior of this hostile body in which he had permitted himself to be trapped. In his weakness he felt himself shrinking, contracting almost to nothingness under the careless pressure of the Master Will.

Clason glanced casually at his watch, and three men converged toward him from as many directions. There was nothing to distinguish them from anyone else in the street, but along the conduits it came to Quest that they were detectives and that they were there by appointment with Keane Clason.

"What floor?" asked the latter, with an excitement which Quest felt instantly was pure pretense. "Are you sure they haven't spirited him away?"

"Don't worry," replied the leader of the detectives. "The alley and roof are covered. We'll take care of the rest ourselves."

On tiptoe they climbed three long flights of stairs in the half-light. Clason held back as if in fear. He was a good actor, and Quest felt the shrinking and hesitation of his body as he crouched and slunk along in the wake of the detectives, pretending terror at what was about to happen, though he knew—and Quest knew he knew—that there would be no resistance up there—that Philip would be found alone exactly as he had been left by Keane's hired thugs.

On the top landing Burke, the leader, paused to count the doors from front to rear.

"This is it," he whispered to the bull-necked fellow just behind him.

The other nodded, and crouched back against the opposite wall while his companions placed themselves in position to cross-fire into the room the moment the door gave way.

Quest longed for the power to kick his hypocrite of a master as he still held back, cowering on the stairs, playing his fake to the limit. Then the door flew in with a splintering shriek under the charge of the human battering ram, and across it hurtled the other two detectives in a cloud of ancient dust.

"Here he is!" someone shouted.

"Phil! Phil!" Keane Clason's voice fairly quavered with sham emotion as he ran into the room and threw himself at a man tightly bound to an upholstered chair, which in turn was wedged in among other articles of stored furniture.

But Philip was too securely gagged to reply, and as Burke slashed the ropes from across his chest he dropped forward in a state of collapse. Stretched on a couch, he soon gave signs of response as a brisk massage began to restore the circulation to his cramped limbs. Suddenly he sat up and thrust his rescuers aside.

"What time is it?" he demanded with an air of alarm.

"One o'clock," replied Keane before anyone else could answer, patting his brother affectionately on the shoulder while within him Quest writhed with indignation. "By Jove! Phil, it's wonderful that we got to you in time. Really, how—you're not injured?"

"No," grunted Philip, "just lamed up. I'll be as fit as ever by to-morrow."

"If you feel equal to it," suggested Burke, "I wish you'd tell me briefly how you arrived here. Do you know the motive behind this affair? Did you recognize any of the body-snatchers?"

Philip frowned and shook his head.

"Yesterday noon," he said slowly, "I took the eight-passenger Airline Express to Cleveland on business. There were three other passengers in the cabin—two men and a woman. Right away I got out a correspondence file and was running over some letters. The next thing I knew I was approaching the ground in the strangest state of mind I ever experienced. My head was splitting, and everything looked unreal to me. Seemed as if I was coming down on some new planet."

"You mean the ship was gliding down to land?"

"No, no. I was dangling from a parachute.... By the way, where am I now?"

"In a Munson Avenue loft."

"In Chicago?"

Burke nodded.

"I guessed as much," frowned Philip. "You see, I came down in a field, and then before I could free myself from my trappings I was pounced on— trussed up and blindfolded—by a gang of men. I knew they had taken me a long distance by automobile, but I saw nothing more until they tore the blindfold from my eyes when they left me here."

"And they were all strangers to you?"

"Yes—those that I saw."

"Isn't this enough for just now, Burke?" interrupted Keane, and Quest received an impression of uneasiness that was not apparent in the inventor's tone. "After a good rest he's sure to recall things that escape him now."

"Just one minute," nodded the detective, turning back to Philip. "Can you think of no plausible reason for this attack? Is there no one who might possibly benefit by putting you temporarily out of the way?"

Philip gave a frightened start. Then he was on his feet, clutching at his brother's arm.

"Keane!" he pleaded, "Keane! What's happened? I know, I know! It's the Projector."

"Water!" roared Keane, and Quest felt the panic that coursed through him as he tried to drown out his brother. "Somebody bring water! He needs it!"

At the same time he snatched up Philip's hand in a grip of steel. Instantly the latter's wild eyes became calm, the flush passed from his relaxing face, and he slumped down weakly on the couch.

In that fleeting moment Quest surged into the body of Philip and confronted his will with a fierce and triumphant ardor. For now his will would have command of a body with which to fight his fiend of a Control.

With a sensation of contempt he met Philip's resistance and buffeted him ruthlessly backward, crushed down and compressed his feebly struggling will. And as Philip yielded, Quest felt his own will expanding to normal, taking possession of the borrowed body with hungry greed, and flashing from its faded eyes the spark of youth.

Burke stared in amazement at the kaleidoscopic rapidity of the changes in the rescued man's expression. Strange lights and shadows continued to flit across Philip's face as Quest's invasion of him proceeded, but with a diminishing frequency which soon assured Keane that his Agent was tightening his command.

The younger of Burke's aides stood fascinated, his mouth agape. The other spoke guardedly to his superior:

"Dope, eh!"

"Nah!" replied Burke, shrugging himself out of his trance. "Shock."

The actual duration of the conflict in Philip was something less than three seconds. It would have been more brief if Quest had exerted himself to the utmost. But his sensations as he first surged into this new habitat under Keane's propulsion were so weird and unearthly that for the moment he was lost in the wonder of the experience. For that short time, therefore, Philip was able to fight back against the onrush of the invading will.

In the next second Quest became conscious of the resistance. Urged on by his Control, he must push Philip back and quell him; but his sympathy for his opponent and his hatred of Keane roused him to sudden revolt. He wanted to disobey the Master Will, retreat, leave Philip in command of himself. But he could only go on, unwillingly thrusting back Philip's will despite the indescribable torment and confusion in his own. Then, with the feeling that he was ten times worse than the most inhuman ghoul, he took full possession of his borrowed body.

"I'll take him home now," said Keane composedly to Burke. "As you see, he needs a little extra sleep. Meanwhile, if you have any occasion to call me, I will be at the factory."

To the youthful mind of the Agent, used to the lightness of an athletic physique, the body in which it moved down the stairs to the limousine seemed strangely heavy and awkward.

"I'm badly done up, Keane," he said with Philip's lips as the car got under way.

"Bah!" snorted Keane, "you've had a scare, that's all. Go to bed when you get home and sleep till nine this evening. At ten a man named Dr.

Nukharin will call for you. He will drive you to a garage, leave the car, and transfer to another one a few blocks away.

"Out near Marbleton you will find an airplane staked in an open field. Nukharin is a capable pilot. He will fly back southeast along the lakeshore to the meeting place. You should arrive about twelve-thirty. The test is set for one o'clock."

Quest listened in a state of abject rage. Lacking the power to resist his Control, he could only boil away in Philip's body like a wild creature hemmed in by bars of steel.

"Bring with you," continued Keane venomously, "the set of papers that you took from the safe in my office. Hold the other set in readiness to deliver to Nukharin to-morrow, after he has studied the results of the test and has notified Paris to release a hundred million dollars in cash for delivery at your Loop office at 3 p. m."

The murderous greed of the man maddened Quest. He tried to revolt, his will squirming like a physical thing, threshing the ether like a wounded shark in the sea. For a moment he felt that he was about to burst the bonds that his demon of a Control had woven around him. So violently did he resist that the immured and sporelike will of Philip forged up fitfully out of the blackness and joined his in the hopeless struggle. But along the attenuated conduits that still chained Quest to the Master Will Keane caught the impulse of the mutiny, and his eyes darted flame as he countered with a will-shock that paralyzed his unruly Agent.

"Listen! you whimpering dog," he snarled. "Think as I tell you—and nothing more! You are going to apologize to Dr. Nukharin for your previous unwillingness to sell the Projector. You are going to tell him that I am at fault—that I held out—but that you found a way to force my compliance. You understand?"

Quest could find no words. With Philip's head he nodded meekly. Just then the car stopped and the chauffeur threw open the door.

Dr. Nukharin flew high despite the masses of cumulus cloud which frequently reduced visibility to zero. He had merely to follow the rim of the lake to his destination, and an occasional glimpse of the water was sufficient to hold him on his course.

In the back seat hunched Philip, his body crumbling under the weight of Quest's despair. For hours the latter had gone on vaguely, hoping somehow to thwart this horrible transaction that was rushing the world to its doom, thinking he might grow strong enough to wrench himself free and so liberate Philip from the dominance of his conscienceless brother. Even though such a move should leave his own will forever separate from his body, he was ready and anxious to make the sacrifice.

Suddenly the crash of the motor ceased and Nukharin banked the ship up in a spiral glide. Quest had never been in the air before, and the long whirl down into the darkness on this devil's errand was to him as eery as a ride to perdition in a white-hot projectile.

His mind seemed to trail out in a great nebular helix behind the descending ship. He felt that he had suddenly crossed some cosmic meridian into a new plane of existence, where he was changed to a gas, yet continued capable of thought. But even here his obsession remained the same. Keane Clason—trickster, traitor, arch-criminal—must be destroyed!

"I'll get him!" vowed Quest in words that were no less real for being soundless. "I'll trail him to the end of space and bring him to account!"

Then wheels touched earth and the cold, bare facts of his destiny rushed in on him with redoubled force. He felt the nearness of his Control seconds before he perceived him through the eyes of Philip. With a sensation like a stab he realized that now he must speak, play his part, be any bloodless hypocrite that Keane Clason chose to make him. The silent order surged down the conduits promptly enough; he responded as an automaton obeys the pressure of a button.

"Well, Doctor," chuckled Philip with a cunning leer, "here's the magic tower, just as I promised you. We'll run it up in a jiffy. This test is going to be so vivid and conclusive that not even a hard-headed skeptic like you can raise a question."

"You misunderstand me," returned Nukharin in an injured tone. "So far as I am concerned this procedure is only a formality, but it is none the less necessary. Suppose that I should spend a hundred million of my government's money and the purchase prove worthless? You may guess that my folly would cost me dear."

Keane Clason was waiting on the platform of a giant truck, the motor of which was idling. All the apparatus was in readiness except that the three demountable sections of the tower had yet to be run up into position.

"One of the beauties of the D. P.," said Philip gleefully to the Doctor, while Keane smiled slyly to himself, "is that this pint-size dynamo provides all the current needed for the test. We pick the power for our radio right out of the air by means of a wave trap and mensurator invented by this bright little brother of mine," and he clapped Keane patronizingly on the back.

"Yes, ah—Dr. Nukharin," ventured Keane timidly, and at that moment Quest experienced the raging red hatred that causes men to murder. "Philip has promised me that you will employ this device only as a threat to hold the ambitions of the larger powers in check."

"Of course, of course!" replied the Doctor heartily. "But now let's have the test. Even at night I'm not too fond of these open-air performances."

The height of the tower as they ran the upper sections into place was forty feet. When all connections had been inspected, first by Keane, then by Philip, the former led Nukharin aloft.

As the climax of his plot approached, Keane's excitement bordered on a cataleptic state, hints of which came confusedly through the conduits to Quest. With a peculiar satisfaction he felt that Keane was suffering. The inventor's jaws became rigid, as though his blood had changed to liquid air and frozen him, and he had difficulty in controlling the movements of his arms.

Now he was afraid! Genuinely afraid, this time. Quest caught the impulse too clearly to doubt its meaning. This was no sham! Keane was doubting his own machine, fearing that in the crisis some element in the finely calculated mechanism might fail to operate, thus cheating him of the blood-money on which his heart was set. Then he was speaking, and even Nukharin noticed the tremor in his voice:

"These nine tubes, which look like a row of gun barrels, are molded from silicon paste. Each shoots a beam of invisible light and a radio dart of precisely the same wave length. The destructive effect depends chiefly upon this exactness of synchronization."

"A question occurs to me," said the Doctor: "will others be able to manipulate the machine as successfully as you can?"

"It's fool-proof," chattered Keane, almost losing control of his voice, "absolutely fool-proof. Surely you have scientists in your country who can follow written directions! Nothing more is necessary."

"Very well," shrugged Nukharin. "I only want to be sure that no unforeseen difficulties may arise in an emergency."

"See this range-setter?" continued Keane. "The thread on the vertical shaft enables us not only to limit the range by angling the beams into the ground, but it can also be disengaged and the Projector revolved in a flat circle for maximum ranges."

"And is there no danger of the machine going wrong—of destroying itself and us?" suggested Nukharin.

"None whatever, Doctor. There is no explosive force and no great electrical voltage involved. As long as we stand back of the muzzles we have nothing to fear.

"Now look. I have set the micrometer at three hundred yards, which will just about cover the stretch between ourselves and the lake. I will cut a swath for you—and every bush, every blade of grass, every insect in this swath will be withered to ash in the twinkling of an eye. The destruction will be absolute."

"Please proceed," said Nukharin grimly.

Keane pulled a lever in its slot, then pressed it down into its lock as his projection battery swung lakeward at the desired angle. Then with one hand poised on another lever, he pressed an electric button.

At the controls below, a bulb flashed on and off. The signal was superfluous, for already Quest had received his silent command from the Master Will. An icy dread fastened on him. He must obey the unspoken command; he had no will of his own with which to resist. The test would be a success; the Projector would be sold; the world would be turned into a shambles. And he, Owen Quest, would be the destroyer, the murderer, the weak fool who made this horror possible.

All this flashed through the Agent's mind in the fraction of a second that it took him to extend Philip's hand, close the switch of the dynamo, and snap on the alternating lights in the housing over the tellurium filter.

For an interminable five seconds he waited, in a ferment of revolt which the paralysis of his will made it impossible to put into action. Then again the command pulsed within him, the signal bulb flashed, and he reversed his motions of the moment before.

Cold sweat cascaded down Philip's face as Quest felt the ladder vibrating under descending feet. He longed for the power to hurl Keane Clason to the ground and turn the Projector upon him. But with an awful irony the Master Will forced him to his feet, and to speak in a tone that withered the manhood within him.

"Come," said Philip in a triumphant tone to Nukharin, "and I will show you that Clason inventions perform as well as they sound."

Flashlight in hand, he started toward the lake with Nukharin and his brother close behind him. Twenty paces, and the long meadow grass suddenly vanished from beneath their feet.

"See that!" whispered Philip excitedly, waving the light from side to side to show the forty-foot swath that stretched away before them. "Not a trace of life left, not a blade of grass—nothing but dust!"

The only response was a gurgling sound that issued from Nukharin's throat.

"Look!" Quest formed the word with Philip's lips under the urge of the Master Will. "Here was a tall bush. What do you see now? Just a teaspoonful of ash. When you examine the remains by daylight, you will find that even the root has disintegrated to a depth of two feet."

"Enough of this," croaked Nukharin in horror. "The deal is closed."

His face was convulsed with fear. Without another word he whirled about and fled toward his airplane. Philip gave a start as if to follow.

"Halt! you slob," growled Keane, whose composure had returned with the successful outcome of the test. "I have use for your company, even though you are as great a coward as our Slavic friend."

Coward! The epithet stung Quest like a flaming goad. One of the fine, intangible lines that bound him under the will of Keane Clason severed, and his own will exploded into action like a thunderbolt. With startling agility he whirled Philip about, the flashlight clubbed in his hand. But Keane was quicker still. A clip on the wrist sent the weapon flying. Then Philip reeled backward from a kick in the stomach, and his clutching hands beat the air as he sank unconscious in the dust.

With a violent tug, Quest lifted Philip's body to a sitting posture. The phone was ringing, and by the pull on the will-fibers he knew that Keane was at the other end of the wire. Philip's body was failing under the strain of the part it was forced to play, and the blow of the night before had further weakened it. Now he sat rocking his head painfully between his hands. But Quest lifted him to his feet by sheer will, and he staggered across the room.

"Hello!", he said in a hoarse voice.

"Get the hell out here to the factory!" rasped Keane, and the crash of the receiver emphasized the command.

It was one o'clock as Philip whirled his sedan into Olmstead Avenue. At three, reflected Quest as the car scorched over the pavements, he must be at the downtown office to deliver the papers and receive the money.

Then he was face to face with Keane, reeling dizzily at the hatred that blazed from the latter's accusing eyes.

"Double-crossed me, eh!" The voice was a low snarl, and as he spoke Keane thumped the extra outspread on his desk. "But you're not going to get away with it—neither of you!"

Dismay, hope, dread, wonder robbed Quest of the power to speak. But he whirled around behind the desk with such unexpected violence that Keane staggered back in alarm. Then he was devouring the screaming headlines of the newspaper. Three seconds, like a slow exposure, and every word of the Record's great scoop was etched upon his mind as if with caustic:

DOOM LAUNCH ADRIFT ON LAKE

Physician Baffled by Condition of
Five Bodies Found in Craft

Blighted Area on Shore Said to
Have Bearing on Tragedy

THAW HARBOR, IND., June 6.—Five Chicago sportsmen, most of them prominent in business and society, perished in the early hours this morning while returning in the launch of A. Gaston Andrews from a weekend camping party near Hook Spit on the Michigan shore.

The boat was towed into this port at daybreak by the Interlake Tug Mordecai after being found adrift less than a mile off shore. According to Captain Goff of the Mordecai the death craft carried no lights and he barely avoided running her down. The weather along the Indiana shore was perfect throughout the night and there is nothing to indicate that the launch was in trouble at any time. The bodies are unmarked, and this little community is agog with rumors ranging all the way from murder and suicide to the supernatural.

Dr. J. M. Addis of Thaw Harbor, the first physician to examine the bodies, says that they appear to have suffered some violent electro-chemical action the nature of which cannot be determined at the moment. This statement is considered significant in view of the reported discovery ashore of a large blighted area almost directly opposite the point where the launch was found. Joseph Sleichert, a farmer who lives in that vicinity, reports that this patch of ground extending back from the lakeshore was completely stripped of vegetation overnight. He ascribes the damage to some unknown insect pest. Others say that the condition of the ground indicates that it has been burned at incinerator temperatures. Nothing is left of the soil but a blue powder.

Philip faced his brother with eyes that were dull with agony.

"You have made me a murderer!" Quest forced out the words in painful gasps.

But Keane snapped back at him like a rabid dog.

"You did it—you did it yourself! You tampered with the Projector. You tried to spoil the test. You changed the range. You tried to kill me, and instead you killed these others. And you're going to pay—both of you. You hear me?—you're going to pay!"

His voice mounted the scale to a scream. It was a wail of unreasoning terror, of the dread of exposure, of the fear that he would fail to collect the fortune now so nearly in his grasp. The accident that had jarred his well-laid plans had unnerved him.

Frantically Quest strove to answer him, to explain his utter subjection, as Agent, to say that if he had possessed the will to oppose or trick him he would have turned him over to the police, or might even have killed him, at the very outset. But in his frenzy, Keane had so tightened his control that Quest was speechless. Now he tried to substitute gesture for words, but Philip was rooted to the spot like a statue; even his hands were immovable.

He might have remained in this state indefinitely had not Keane's fears withdrawn his mind from his immediate surroundings. Momentarily he forgot Quest, Philip—everything but himself and his predicament. And in the instant that his vigilance relaxed, Quest's enslaved will experienced a sudden lease of strength and hope. Independently of his Control, he found that he could move Philip's hand, could take a faltering step.

But now, what to do? How might he fan this feeble spark of volition to sufficient strength for decisive resistance? The idea came to him: if only he could place distance between himself and Keane, perhaps with one titanic effort he might launch himself against the Master Will, take him by surprise, crush him down, and reverse him to the status of Agent instead of Control.

With infinite effort Quest forced Philip's body step by step across the room. He must reach that window, get a signal of distress to someone in the street.

But Keane began to sense a mutiny. He followed. He crossed the floor with slinking, tigerish steps and snaking body. His wet lips writhed back over his teeth, and his contorted features wove the leer of the abyss. Now as his Control drew physically near, Quest felt his mite of strength ebbing fast. Slowly Keane reached up with his clawed fingers and grasped his Agent by the arm.

"Remember!" he hissed, "if these deaths are traced to us, you break down—you confess—you take the blame—you paint me lily white—you describe the cowardly means by which you moulded me to your will—you plead only for a quick trial and the full penalty of the law. You understand?"

Quest made no reply, but he understood all too well the hideous intention of his betrayer. What a fool he had been to imagine that Keane Clason would ever restore him to his body! Philip to the chair, Quest a homeless spirit wandering in space, and for the body at the bottom of the tank, the brief regrets of the Department!

A sudden rushing sound filled the air with a sense of action and alarm.

Two—three—four speeding automobiles swung in recklessly to the curb and shrieked to a standstill under smoking brakes. Men leaped out and deployed on the run to surround the factory. Keane darted to the door and twisted the key.

"Come on!" he spat at Philip as he snatched back the rug and threw open the trap door.

The command galvanized Quest to action. In two bounds he had Philip on the stairs. A heavy impact rattled the office door just as he dropped the trap into place over his head. Then, infected with Keane's panic, he was running down the passageway like mad.

Inside the tank chamber the brilliantly colored rings of liquid flashed back the rays of the arclight. Half crazed with anxiety, Keane danced on the black ledge like a monkey on a griddle. His face was ashen, drool ran from his twisted mouth, his eyes were two black pools of terror.

Again Quest experienced the peculiar sensation which came with the slackening of control. New hope sprang up in his agonized being as heavy blows boomed against the air-locked door. Great waves of fear poured along the conduits, betraying to the Agent the state of mind of his Control. Now what would Keane do? What could he do? Why, of all places, had he fled down into this blind burrow?

Thud, thud! Then came a series of sharp reports. Outside, they were trying to shoot away the deep-sunk disk hinges.

Still the door stood fast, but the fury of the assault on it whipped the faltering Keane to action. In a bound he was on the platform. With a lightning hand he threw the switch to plus, starting electrolytic action in the tank. Then he pressed a button concealed under the edge of the switch-mount and a panel slid silently aside in the wall, revealing a narrow outlet.

To Quest everything went a flaming red. He might have known that this fox would have something in reserve—a way of escape when danger threatened!

But his Control gave him no time for independent thought. He forced Quest to turn Philip's eyes up to his own. Without disconnecting that grip of his glittering eyes, Keane leaped back to the ledge. Quest felt the silent order:

"Get up on that plank! Dive into the tank! Get back into your own body, let Philip have his! Then come up—the two of you—and face the music. For I'll be gone, and your story will sound like the ravings of a maniac."

Quest took an obedient step toward the platform. But at the same instant a tremendous crash shivered the door. It seemed to unnerve Keane Clason. With a gasp he sank down upon the steps, his body doubled in pain, his hand clutching at his heart. Another crash followed, and he shuddered and cried out.

Instantly Quest felt an expansion of the will. Keane's sudden physical weakness had loosened his control. Philip's lips worked painfully as Quest forced him to pause, to disobey the command of the Master Will. In a spasm of will he fought to wrench himself free from the countless clinging tentacles of his Control. In great surges, Quest's reviving volition pounded against the walls of his borrowed body. Now he sought to force this sluggish body back to the wall, so that he might release the airlock and spring the door. But Philip seemed to ossify, every cord and muscle of his body frozen to stone by the conflict that raged within him.

Braced against the wall, Keane was rising slowly to his feet. His seizure was easing, and so he was able to exert a better pressure upon his rebellious Agent.

"Come!" he gasped, realizing that he lacked the strength to escape alone and must therefore change his plan. "Lift me—quick! Carry me out! Slide the panel back into place. We will escape together!"

The spoken command turned the balance against Quest. His will yielded to the master. At the same instant Philip's body relaxed like an object relieved of a great excess of electrical potential. Suddenly strong and supple, he lifted the trembling Keane and tossed him across his shoulder.

For a moment there had been a lull in the assault on the door. Now the battering resumed with a fury that jarred the whole chamber and sent ripples dancing across the varicolored liquids in the osmotic tank.

"Quick!" gasped Keane. "Move! I say. Carry me out."

But he was in a fainting condition. Crash after crash rocked the chamber, and with every blow Quest's will felt a stimulation that enabled him to stand off the commands of his Control. Then a wave of nausea swept over him and left him reeling. It seemed that Philip's blood had turned to boiling oil. A dazzling mist swallowed him up, and with a weird sense of inflation he felt full strength returning to his will.

A booming blow that bulged the door inward acted upon him like a stage player's cue. He leaped to the platform. The gurgling sound of remonstrance rattled from Keane's throat. But Quest paid no heed. Philip was walking the plank—away from the open panel—out over the tank.

Rapidly he dropped down the ladder to the bottom rung, snatched Keane's wrist in a gorillalike grip, and hurled him down into the vat.

Then Philip was clinging desperately to the ladder, his strength gone, his body shivering as if with ague.

"Go on up!" came a strange, impatient voice from below him. "For heaven's sake let me out of here!"

A downward glance, and with a shout of alarm Philip was scrambling up the ladder, for there was a head down there, and a pair of naked shoulders, and the face of a man he had never seen before. Hand over hand Quest followed. Philip had collapsed and lay prone on the plank. Quest lifted him to his feet and shook him anxiously.

"Philip!" he urged. "Philip! Can you walk?"

The tattoo on the battered door helped to revive the older man.

"Quick!" whispered Quest, kneading Philip's arms. "There's barely an hour left. Get to your office. Burn the papers. Refuse the money. Do you hear me?"

Philip nodded dazedly.

"Hurry!" puffed Quest, thrusting him through the opening that Keane had reserved for his own escape, and sliding the panel back into place.

Quest was himself now—young, strong, free. Instantly he threw the electrolytic switch to minus. For Keane had failed to emerge from the tank, and since he was submerged alone, he could not escape until electrolysis was halted.

Just as Quest leaped from the platform to release the airlock, the door burst in and three men with drawn guns rushed into the chamber.

The leader stopped with a startled oath and stood blinking his unbelieving eyes. Quest was poised like a statue, his naked body gleaming an unearthly white against the lusterless black of the wall.

"Quest," came from the three in chorus. Then a rush of questions: "What's the matter? What's happened to you? Where are the Clasons?"

Quest turned toward the platform, expecting to see Keane.

"Something's wrong!" he shouted. "Quick! Somebody get Philip. He's gone to his Loop office. Keane Clason's at the bottom of this tank. I'm not sure how this thing works, but Philip can get him out! I'm sure of it!"

Despite the confident predictions of both Quest and Philip Clason, osmotic association failed to restore Keane to life, and at last the coroner ordered the removal of the body. The autopsy revealed heart disease as the cause of his death.

For reasons best understood at Washington, the cause of the five launch deaths was withheld from the public. Quest's punishment for his part in the crime consisted of a promotion and a warm personal letter from the President of the United States.

Compensation

By C. V. Tench

Professor Wroxton had disappeared—but in the bottom of the mysterious crystal cage lay the diamond from his ring.

"Why, John!" Involuntarily I halted at the entrance to my snug bachelor quarters as the flood of light my turning of the switch produced revealed a huddled figure slumped in an easy chair.

"Aye, sir, 'tis me." The man got to his feet, gnarled hands rubbing at his eyes. "An' 'tis all day that I've been waiting for you, sir. The caretaker said you'd be back soon so let me in. I must have fell asleep, an' no wonder, what with the strain an' no sleep or rest all last night."

"Strain? No rest?" I stared my bewilderment, trying at the same time to conceal the vague apprehensions occasioned by the fact that the trusted servitor of my friend, Professor Wroxton, should wait all day for me.

Hastily shedding my outer things, I bade him again be seated, sat down facing him, and asked him to explain.

"'Tis the professor, sir." The old chap peered at me with anxious, wrinkled eyes. "'Tis common enough for him to send me here on messages, sir, but to-day I've come on my own, because, sir," answering the question in my eyes, "I haven't seen sight of him since last night."

"Why—" I began.

"That's just it, sir." John took the words out of my mouth. "For twenty years my wife an' me have looked after the professor at The Grange. In all that time he's never been away at night. Whenever he had to come to town he'd tell us. Most times I'd drive him myself in the old car. But that was very seldom, sir, for Professor Wroxton had few interests outside."

"But, John," I protested "is there no other reason for your agitation? He might have had an urgent call, or gone out for a walk or drive by himself."

"No, sir. If you'll pardon me, sir, you're wrong. The professor was fixed in his habits. He would not go away without tellin' me. Think back, sir, you know the professor as well as me. Better, because you are his friend and I am only a servant. Although, sir," this proudly, "he always treated me as a friend."

"Go on," I urged, seeing he was not finished.

"Well, sir, a few minutes back you asked me if there was no other reason for my being upset like. There is, sir. You know, sir, that for more'n twenty years the professor has led a retired sort of life; the life of a—a—"

"Recluse," I suggested.

"That's it, sir. He only left The Grange when he had to. He was all wrapped up in some weird-like thing he was inventing. In all those years, sir, you were the only visitor who ever went into his laboratory, or stayed at The Grange for a night or more. That is, sir, until three days ago."

"Go on," I again urged, some of his perturbation communicating itself to me.

"The Grange, sir, lying as it does, fifteen miles from town an' back in its own grounds away from the road, isn't noted by many. When strangers do get into the grounds I usually gets 'em out again in short order. Three days ago, sir, a stranger drove up to the door in a fine car. He told me he was wantin' to purchase a country home. I told him The Grange was not for sale an' turned 'im away. He was turning his car to leave when my master came out. To my surprise, sir, he invited the stranger in. An' I'm sure, sir, because he looked so taken aback like, that the stranger had never seen the professor before."

"And after that?" I asked, now feeling decidedly uneasy.

"The stranger, sir—a Mr. Lathom he called himself—stayed on. He was in the study with the master last night. This morning there was no trace of either of them."

"But—good God, John!" I jerked to my feet, a fresh dread clutching at my heart. "What are you trying to get at? The professor and Mr. Lathom might possibly have driven away somewhere last night."

"Both cars, sir," the servant answered, "are in the garage. I bolt all the doors in the house myself every night. They were still fastened this morning. My wife an' me searched the house from cellar to garret an' hunted all over the grounds. We couldn't find a trace of the master or his guest."

"You mean to suggest then," I shot at him, "that two full grown men have completely vanished? It's absurd, John, absurd!"

I paced the floor thinking desperately for a few minutes, conscious of the ancient's anxious eyes. I half smiled. The thing was too ridiculous for anything. Old John had grown morbid from living away from the outer world. Also, I had to admit that the atmosphere of The Grange, impregnated as it was with the lethal scientific dabblings of my friend, was exactly suited to the conjuring up of unhealthy forebodings in uneducated minds. I'd drive out to the home of my friend at once. No doubt I'd find him fit and well. He had refused to install a phone, so drive it had to be.

"John." I stopped my pacing and patted him on the shoulder. "I'm coming out to The Grange at once." His face showed his thankfulness. "I am sure," I went on as I struggled into my coat, "that we shall find the professor and his guest awaiting us. Anyway, it's time you got back to your wife and had some food."

"I hope to Heaven, sir, that you're right." With that we left the building and entered my car.

Although I had tried to dispel my fears, although I had tried to banter John out of his dread, I drove that evening as I had never driven before or since. Barely fifteen minutes later I halted my roadster at the short flight of steps leading to the main door of The Grange. Even as we stepped from the machine the door flung open and an agitated woman hurried towards us. She was Mary, John's wife.

"Sir!" She gripped my arm and stared anxiously into my face. "'Tis glad I am that you've come. The Grange is a house of death."

In spite of myself a chill shook my whole body. Gently handing her to John, I strode up the steps.

At the open doorway I halted, the aged couple crowding on my heels, the woman still babbling about death. I couldn't blame her. All day she had been alone in that gloomy, rambling old building, wondering, no doubt, why John and I had not returned sooner.

And gloomy the house was. Always, even when staying there at the professor's request, I had found it to be somber and depressing, as if there lurked within its walls the shadowy wings of the years-old tragedy that had caused my friend to retire to such a God-forsaken place, and there become absorbed in his scientific experiments.

Even now, as I gazed into the dimly-lighted hallway, the air seemed charged with that same malignant something I cannot describe.

Pulling myself together I strode quickly along the corridor, and flung open the study door. The lights being full on, one glance sufficed to show me that my friend was not there. Swinging on my heel, the horror I saw in the eyes of the servants, honest, healthy folks not easily frightened, conveyed itself to me. Somehow, the sight of that room, lights on, chairs drawn up to the burnt-out fire, brought home to me the fact that something serious was amiss. I chided myself for thinking John had been unduly agitated.

For a moment I stood, trying to conceal the chill coursing through my veins, puzzling what to do next. I decided to search the house thoroughly. If I found no sign of the professor or his guest, I would call in the police.

Fearfully yet willingly the aged couple led me from room to room, from attic to basement, until but one place remained—the laboratory. I hesitated for several seconds at the closed door of my friend's workroom. Not that I had never entered the—to a layman's eyes—weirdly-appointed place. I had been in many times with the professor. But this time I dreaded what I might find.

Pulling myself together, I gently tried the door. To my horror it yielded to my touch. Alive, the professor always kept it locked. A new dread assailed me, as, flinging the door wide open, I blinked in the sudden glare of powerful globes. Someone had left the lights full on!

Horrified I stood and stared, knowing by their heavy breathing that the aged couple were also staring with fright-widened eyes. Afraid of what? I did not know. I only knew that the atmosphere had become even more sinister. I knew that something dreadful had taken place in that room.

Trembling with consternation I forced myself to take a few steps forward, then I again stared about me. At one end of the large room something shone brightly in the glow of the lights. Slowly I walked across to examine it: it appeared to be a glass case, almost like a show-case, about eight feet square and seven feet in height. With the mechanical actions of the mentally distraught I walked all around it. Not the slightest sign of an entrance could I see. The fact intrigued me. I tapped lightly on the highly polished surface with my fingers. It rang to my touch like cut glass.

Through the transparent surface I could see John and his wife. They were watching me furtively, wondering, no doubt, why I lingered. As I looked at them John suddenly lumbered up to the case on the opposite side. Dropping to his knees, he stared. Turning an imploring gaze to me, he pointed. His lips moved soundlessly. I followed the pointing finger with my eyes; gasped at what I saw.

Near the center of the cage, on the floor constructed of the same crystalline substance, something glittered, its brilliance almost dazzling as the light rays struck it. My face pressed close to the cold outer surface of the structure, my shocked intelligence gradually realized what that small sparkling object was. It was a magnificent diamond—and the professor had always worn a diamond ring!

In a sudden frenzy of horror I pawed my way around the cage to where John still knelt. As I reached him he jerked his head in a numb way as he croaked, "It's a diamond, sir! The professor's!"

"But how?" I implored. "How can it be? There's no way into this thing. Perhaps he was working here, and the stone came loose from its setting. He couldn't have dropped it after the cage was completed."

"It's his diamond, sir," intoned the old man, dully. "I know it is."

Then a sudden unreasoning terror filled me. I shrank away from that shining box. It seemed to be mocking me, gloatingly, malevolently.

"Quickly!" I threw at the aged couple. "Let us get out of here! Now! At once!" They needed no second urging. I knew that they felt as I felt: the laboratory was a sepulcher!

Five minutes later I was guiding my car over the narrow road to town. I did not pause until I drew up at police headquarters. I suppose my appearance was distraught, for I was ushered into the presence of the chief without delay. In a few moments I had poured out my story. He listened with a polite calmness I found almost maddening. Leaning back in his chair, he reviewed, audibly, the facts.

"Some twenty-odd years ago your friend, Professor Wroxton, married. He was so absorbed in the pursuit of some weird invention that he neglected his bride. She ran away with another man. This man deserted her, and disappeared. The professor found her many months later, in desperate health. Shortly afterwards she died. Your friend tried to trail the man, but failed. Shocked and saddened beyond measure, he retired to a place known as The Grange."

He suddenly straightened up in his seat, and pointed at me a thick forefinger.

"How long have you known Professor Wroxton?"

"About ten years," I answered.

"What was he trying to invent?"

"I don't know," I replied.

"And yet you had his confidence in other matters?"

"But what has all this to do with finding out what has become of my friend?" I blurted out. "Perhaps every moment counts."

"A lot." The chief eyed me in a way I did not like. "Solely because your friend has not been seen by his servants for nearly twenty-four hours, merely because you saw what you believe to be his diamond in some kind of a glass compartment in his laboratory, you come here as distraught as a man who has something terrible on his mind. Why?"

"I can't say." I shifted uneasily under that direct stare. "Somehow I *feel* that something dreadful has happened to my friend."

"We do not go by *feelings*." The chief got to his feet. "But you have told me enough to warrant action. I want you to guide me and a couple of men to this house. Please wait here until I return." He left the room.

Sitting there awaiting his return, I tried to ponder the matter reasonably. After all, perhaps the chief was right. Merely because the professor had been absent for a few hours and I had seen what I thought to be his diamond in the laboratory, I had worked myself into a perfect fever of anxiety. I almost smiled to myself. In that businesslike office the whole affair did seem absurd. After all the professor did not have to answer to his servants for his actions.

Heavy footsteps, announcing the chief's return, caused me to rise to my feet. A few minutes later, in company with the three officers, I was driving again towards The Grange.

We made the return journey in almost complete silence. Occasionally the chief would shoot a question at me; but, the night air cooling my fevered brain, my replies were guarded. He realized that fact, for I felt his eyes upon me all the way. What was going on behind that broad forehead, I wondered.

Then we reached The Grange. As we mounted the steps, John, his wife herding behind him, flung wide the door. He answered the question in my eyes with a negative shake of his head, and the words, "Nothing fresh, sir."

The chief eyed him keenly, then curtly bade him lead the way to the laboratory. John hung back, his face blanched. "I can't, sir," he faltered. The chief turned to me, and, although I wanted to follow John's example, although the atmosphere of the house had again filled me with an unshakable dread, I led the way, standing back at the door to allow the officers to enter first.

With calculating gaze the chief slowly took in every detail of the stone apartment. He turned to me.

"What is there here to be afraid of?" I pointed hesitatingly towards the crystalline cage. The chief and his men strode across to it.

"You don't know how to open this?" the chief shot at me after a brief examination.

"No," I replied. "It was not here on my last visit."

"When was that?"

"Some two or three months ago", I answered. "My work occasions much traveling on my part."

The chief and his men turned again to the cage, talking in undertones. He turned again to me.

"You notice that this thing is built in sections. One of them must be movable. Perhaps—" He paused as his eyes fell upon some wires and tubes that trailed across the floor from underneath the cage to a switchboard fastened to the wall.

"Perhaps," he repeated, "it is worked from that board." He crossed over, stared thoughtfully at the shining levers for some seconds, and moved one slightly. The result was astounding. All four of us stared with unbelieving eyes as slowly, without the faintest sound, a section of one wall slid inwards, as if guided by invisible tracks on floor and ceiling.

"Guess that's enough for now." With the words the chief backed away, almost timidly, I thought, from the switchboard, and walked to the cage. For a moment he hesitated, but he entered, and emerged with the sparkling object in his hand.

"It's the professor's," I choked, crowding close to him.

"How'd you know?" he shot back. "All unset stones look pretty much alike."

"I just know," was all I could falter.

"You 'just know'." The chief sat down on a stool and regarded me searchingly. "Mr. Thornton, when I started out with you, I thought I was on a wild goose chase or the trail of a confession. You looked exactly like a man who had either committed a serious crime, or was getting over a bad drunk. I feel sure now"—he again regarded the diamond—"that your story was not the product of an alcohol-crazed brain. Come on!" He lurched to his feet, and grasped me by the shoulder. "Come through!"

Without answering, I wrenched myself free. Over my shoulder I saw one of the policemen at the door. In the hand of the other a revolver suddenly appeared. Good God! I glared in bewilderment from one to another. Was I going mad? Surely this was some awful nightmare! What had I said to make them suspect me of having committed a revolting crime?

"Sit down!" The command came from the chief. Mechanically I found a stool, and obeyed him. "Hold your stations, boys, and listen carefully," he ordered his men. Then he turned to me.

"Professor Wroxton was a wealthy man without kith or kin?"

"Yes."

"Do you know the nature of his will?"

"Yes." Chilled to the heart, I felt the circumstantial net tightening.

"What is its nature?"

"This house and an annuity to John and his wife," I explained. "The residue of his wealth to me."

"Humph!" The chief stared at me piercingly. "And how has business been with you lately?"

Damn the man! What right had he to put me through the third degree? I felt my state of dazed horror slowly giving way to anger. I glanced around. The pistol still menaced; the man at the door had not moved. It was useless to try and evade the questions.

"For the past year," I replied, "business has been very poor. In fact, the professor advanced me some money."

"Humph!" Again that irritating, non-committal grunt.

The chief turned in his seat and stared thoughtfully at the crystalline cage.

"And you don't know what the professor was trying to invent?"

"Only its nature," I began.

"Ah! That's better. Why didn't you tell me that before?" The chief leaned forward.

"Well," I explained, "the whole thing seems so absurd. When the professor told me how his married life had been broken up, he told me that at that time he reached the utmost depths of human suffering. Absolute zero, he called it."

"Ah!"

"The experiments he indulged in," I continued, trying to hide the shiver pimpling my flesh, "were to produce an actual state of absolute zero. It is years since he told me this. I had almost forgotten it."

"And exactly what is an absolute zero?" The chief's eyes never left mine.

"Well," I protested, "please understand that I also am a layman in these matters. According to my friend, an absolute zero has been the dream of scientists for ages. Once upon a time it was attained, but the secret became lost."

"And exactly what is an absolute zero?"

Curse the man! I could have struck him down for the chilling level of his tone. I forced myself to go on, realizing that I was damning myself at every step.

"An absolute zero is a cold so intense it will destroy flesh, bone and tissue. Remove them," my voice rose in spite of myself, "leaving absolutely no trace."

No trace! Something attracted my eyes. The chief had opened his hand. The diamond there flashed and sparkled as if mocking me. I pulled myself together, and went on.

"It all comes back to me now. One day I came out here and found the professor terribly distraught. He told me that, with the aid of electric currents he had been able to invent the absolute zero, but he could not invent a *container*."

"Why?" Those eyes continued to bore into mine.

"Because—remember it is years since he told me this—there was difficulty in controlling the power. Besides destroying living things, it would destroy bricks and mortar, stone and iron. Only one substance it could not wipe out—crystalline of diamond hardness.

"I know, now!" I jumped to my feet and grabbed the chief's arm. "I know now what he meant. Fool, fool! Why did I not think of it before? This—" I swung towards the cage—"is compensation." Almost panting in my eagerness I went on:

"My friend told me that the law of compensation would atone to him for the tragedy of his youth. Absolute zero in suffering would be atoned for by a real state of absolute zero. Chief!" I whirled on him. "Don't you understand? This is the perfected dream of my friend. It is the absolute zero."

"Humph! Plausible but not convincing." I slumped back at the officer's words. "That does not explain the professor's disappearance. Even if it did, what about Mr. Lathom? And don't forget this contrivance is worked from outside. We found the diamond inside. Of course, he might have placed it there himself to test the machine," he concluded.

"Of course, that's it," I commenced. But I regretted the words when I saw suspicion flicker again in the chief's eyes. Lamely I finished, "And he has probably rushed off, in an ecstasy of triumph, to acquaint professional colleagues."

"Without unlocking any doors or taking a car, eh?

"Mr. Thornton." The chief stood up and regarded me sternly. "As a sensible man, don't you think yourself that your story is a bit thin? The professor has disappeared. Here is a strange-looking case which you say is an absolute zero container. Whether you know, or are just jumping at conclusions, remains to be proved. But even if it is, do you think that, after perfecting such a tremendous invention, the professor would commit suicide?"

"On the contrary," I gasped, "my friend was a man of gentle, kindly disposition, but strong purpose. I should think his first action on attaining his life's ambition would be to notify me, his closest friend."

"And he didn't." Every word condemned me, and roused me to retaliate.

"Chief, I know enough of the law to know that, before you can try a man for murder, you must prove that murder has been committed." I grinned savagely. "You must have the corpus delicti. Go ahead! Find my friend or his remains, or else withdraw your charges." I grinned again, with shocked mirthlessness.

Then I buried my head in my hands. I had called in the police to help find the professor, and they had only blundered around and asked a lot of stupid questions. The chief had practically accused me of murder—something I knew he could not prove, yet feared he might. Because I had told the chief of the locked doors and unused cars, he had confined his investigations to the house itself.

He interrupted my thoughts.

"Mr. Thornton, I am going back to town. You will remain here with my men. I advise you to get some sleep, as I shall not be able to carry out certain investigations until the morning. One of my men will spend his time searching the house and patrolling the grounds, the other one will stay here with you."

He turned away, whispered some instructions to his men, and, followed by one of them, silently left the laboratory. I started to protest, tried to follow him; the man at the door stopped me. Silently, almost grimly, he indicated a narrow cot at one end of the room. For a moment I hesitated, feeling the man's eyes upon me.

Sleep on my dead—I felt sure he was dead—friend's cot! Sleep in that fearful place! My whole being crawled with horror. I turned again to the man. His features were unyielding. Perhaps this was more third degree. Limp with weakness and weariness, I dragged my lagging feet towards the cot.

As long as I live I shall never forget my awakening. A uniformed figure, the chief, shaking me by the shoulder. Two other uniformed men silently watching. I sat up and gazed about me, dazedly. Bright sunlight streamed through the windows. A stray gleam struck the cage. I shrank back, trembling. And yet I had slept soundly.

"Mr. Thornton," the chief said, "I have serious news for you. I have positive proof your friend is dead."

"Dear God!" The exclamation was wrung from me as recollection returned with a rush. "Where? You can't have!"

"Here." He thrust a bundle of letters into my hands. "You acted so strangely last night you caused me to suspect you of a serious crime. Also, you overlooked several important points. You got back from a trip only last night."

Last night! Surely it was years.

"You had left instructions to have your mail forwarded," the level voice went on. "These letters were evidently one day behind you. I picked them up at your rooms this morning. I took the liberty of opening them. Read this one." He selected it.

With trembling fingers I extracted from the envelope a single written page. I recognized the handwriting as the professor's. I read with feverish intensity, each single word burning itself into my consciousness:

Dear Thornton:

I am writing this in anticipation. I will see that it is mailed when my plans are completed. Too late, dear friend, for you to attempt, with the best intentions in the world, to frustrate them.

You will, perhaps, recall that many years ago, when I gave you my full confidence, I told you that I felt sure that the law of compensation would atone in some measure for my loss. Thornton, old friend, I believe that, in more ways than one, my hour has arrived. Two days ago I completed the absolute zero. But even better!

A man called here to-day. Although he did not recognize me, I saw through the veneer of added years with ease. Fate, call it what you will, my visitor is the man who wrecked my happiness.

Under pretext I shall detain him. I shall induce him to enter the crystalline cage. I have already arranged a dual control which the power will destroy when I apply it from *the inside of the cage.*

Please destroy the cage. It will have brought compensation to me before you read this.

Good-by, dear friend!

Wroxton.

"I apologize, Mr. Thornton." The chief offered a hand which I clutched in mingled sorrow and relief. The world had lost a genius. I had lost a dear friend. But he was right. It was compensation.

Tanks

By Murray Leinster

... The deciding battle of the War of 1932 was the first in which the use of infantry was practically discontinued ...

—History of the U.S., 1920-1945 (Gregg-Harley).

Two miles of American front had gone dead. And on two lone infantrymen, lost in the menace of the fog-gas and the tanks, depended the outcome of the war of 1932.

The persistent, oily smell of fog-gas was everywhere, even in the little pill-box. Outside, all the world was blotted out by the thick gray mist that went rolling slowly across country with the breeze. The noises that came through it were curiously muted—fog-gas mutes all noises somewhat—but somewhere to the right artillery was pounding something with H E shell, and there were those little spitting under-current explosions that told of tanks in action. To the right there was a distant rolling of machine-gun fire. In between was an utter, solemn silence.

Sergeant Coffee, disreputable to look at and disrespectful of mien, was sprawling over one of the gunners' seats and talking into a field telephone while mud dripped from him. Corporal Wallis, equally muddy and still more disreputable, was painstakingly manufacturing one complete cigarette from the pinched-out butts of four others. Both were rifle-infantry. Neither had any right or reason to be occupying a definitely machine-gun-section post. The fact that the machine-gun crew was all dead did not seem to make much difference to sector H.Q. at the other end of the telephone wire, judging from the questions that were being asked.

"I tell you," drawled Sergeant Coffee, "they're dead.... Yeah, all dead. Just as dead as when I told you the firs' time, maybe even deader.... Gas, o'course. I don't know what kind.... Yeh. They got their masks on."

He waited, looking speculatively at the cigarette Corporal Wallis had in manufacture. It began to look imposing. Corporal Wallis regarded it affectionately. Sergeant Coffee put his hand over the mouthpiece, and looked intently at his companion.

"Gimme a drag o' that, Pete," he suggested. "I'll slip y' some butts in a minute."

Corporal Wallis nodded, and proceeded to light the cigarette with infinite artistry. He puffed delicately upon it, inhaled it with the care a man learns when he has just so much tobacco and never expects to get any more, and reluctantly handed it to Sergeant Coffee.

Sergeant Coffee emptied his lungs in a sigh of anticipation. He put the cigarette to his lips. It burned brightly as he drew upon it. Its tip became brighter and brighter until it was white-hot, and the paper crackled as the line of fire crept up the tube.

"Hey!" said Corporal Wallis in alarm.

Sergeant Coffee waved him aside, and his chest expanded to the fullest limit of his blouse. When his lungs could hold no more he ceased to draw, grandly returned about one-fourth of the cigarette to Corporal Wallis, and blew out a cloud of smoke in small driblets until he had to gasp for breath.

"When y' ain't got much time," said Sergeant Coffee amiably, "that's a quick smoke."

Corporal Wallis regarded the ruins of his cigarette with a woeful air.

"Hell!" said Corporal Wallis gloomily. But he smoked what was left.

"Yeah," said Sergeant Coffee suddenly, into the field telephone, "I'm still here, an' they're still dead.... Listen, Mr. Officer, I got me a black eye an' numerous contusions. Also my gas-mask is busted. I called y' up to do y' a favor. I aim to head for distant parts.... Hell's bells! Ain't there anybody else in the army—" He stopped, and resentment died out in wide-eyed amazement. "Yeh.... Yeh.... Yeh.... I gotcha, Loot. A'right, I'll see what I c'n do. Yeh.... Wish y'd see my insurance gets paid. Yeh."

He hung up, gloomily, and turned to Corporal Wallis.

"We' got to be heroes," he announced bitterly. "Sit out here in th' stinkin' fog an' wait for a tank t' come along an' wipe us out. We' the only listenin' post in two miles of front. That new gas o' theirs wiped out all the rest without report."

He surveyed the crumpled figures, which had been the original occupants of the pill-box. They wore the same uniform as himself and when he took the gas-mask off of one of them the man's face was strangely peaceful.

"Hell of a war," said Sergeant Coffee bitterly. "Here our gang gets wiped out by a helicopter. I ain't seen sunlight in a week, an' I got just four butts left. Lucky I started savin' 'em." He rummaged shrewdly. "This guy's got half a sack o' makin's. Say, that was Loot'n't Madison on the line, then. Transferred from our gang a coupla months back. They cut him in the line to listen in on me an' make sure I was who I said I was. He recognized my voice."

Corporal Wallis, after smoking to the last and ultimate puff, pinched out his cigarette and put the fragments of a butt back in his pocket.

"What we got to do?" he asked, watching as Sergeant Coffee divided the treasure-trove into two scrupulously exact portions.

"Nothin'," said Coffee bitterly, "except find out how this gang got wiped out, an' a few little things like that. Half th' front line is in th' air, the planes can't see anything, o'course, an' nobody dares cut th' fog-gas to look. He didn't say much, but he said for Gawd's sake find out somethin'."

Corporal Wallis gloated over one-fourth of a sack of tobacco and stowed it away.

"Th' infantry always gets th' dirty end of the stick," he said gloomily. "I'm goin' to roll me a whole one, pre-war, an' smoke it, presently."

"Hell yes," said Coffee. He examined his gas-mask from force of habit before stepping out into the fog once more, then contemptuously threw it aside. "Gas-masks, hell! Ain't worth havin'. Come on."

Corporal Wallis followed as he emerged from the little round cone of the pill-box.

The gray mist that was fog-gas hung over everything. There was a definite breeze blowing, but the mist was so dense that it did not seem to move. It was far enough from the fog-flares for the last least trace of striation to have vanished. Fifteen miles to the north the fog-flares were placed, ranged by hundreds and by thousands, burning one after another as the fog service set them off, and sending out their incredible masses of thick gray vapor in long threads that spread out before the wind, coalesced, and made a smoke-screen to which the puny efforts of the last war—the war that was to make the world safe for democracy—were as nothing.

Here, fifteen miles down wind from the flares, it was possible to see clearly in a circle approximately five feet in diameter. At the edge of that circle outlines began to blur. At ten feet all shapes were the faintest of bulks, the dimmest of outlines. At fifteen feet all was invisible, hidden behind a screen of mist.

"Cast around," said Coffee gloomily. "Maybe we'll find a shell, or tracks of a tank or somethin' that chucked the gas here."

It was rather ludicrous to go searching for anything in that mass of vapor. At three yards distance they could make each other out as dim outlines, no more. But it did not even occur to them to deplore the mist. The war which had already been christened, by the politicians at home, the last war, was always fought in a mist. Infantry could not stand against tanks, tanks could not live under aircraft-directed artillery fire—not when forty guns fired salvos for the aircraft to spot—and neither artillery nor aircraft could take any advantage of a victory which either, under special conditions, might win. The general staffs of both the United States and the prominent nation—let us say the Yellow Empire—at war with it had come to a single conclusion. Tanks or infantry were needed for the use of victories. Infantry could be destroyed by tanks. But tanks could be hidden from aerial spotters by smoke-screens.

The result was fog-gas, which was being used by both sides in the most modern fashion when, their own unit wiped out and themselves wandering aimlessly in the general direction of the American rear, Sergeant Coffee and Corporal Wallis stumbled upon an American pill-box with its small garrison lying dead. For forty miles in one direction and perhaps thirty in the other, the vapor lay upon the earth. It was being blown by the wind, of course, but it was sufficiently heavier than air to cling to the ground level, and the industries of two nations were straining every nerve to supply the demands of their respective armies for its material.

The fog-bank was nowhere less than a hundred feet thick—a cloud of impalpable particles impenetrable to any eye or any camera, however shrewdly filtered. And under that mattress of pale opacity the tanks crawled heavily. They lurched and rumbled upon their deadly errands, uncouth and barbarous, listening for each other by a myriad of devices, locked in desperate, short-range conflict when they came upon each other, and emitting clouds of deadly vapor, against which gas-masks were no protection, when they came upon opposing infantry.

The infantrymen, though, were few. Their principal purpose was the reporting of the approach or passage of tanks, and trenches were of no service to them. They occupied unarmed little listening-posts with field telephones, small wireless or ground buzzer sets for reporting the enemy before he overwhelmed them. They held small pill-boxes, fitted with anti-tank guns which sometimes—if rarely—managed to get home a shell, aimed largely by sound, before the tank rolled over gun and gunners alike.

And now Sergeant Coffee and Corporal Wallis groped about in that blinding mist. There had been two systems of listening-posts hidden in it, each of admittedly little fighting value, but each one deep and composed of an infinity of little pin-point posts where two or three men were stationed. The American posts, by their reports, had assured the command that all enemy tanks were on the other side of a certain definite line. Their own tanks, receiving recognition signals, passed and repassed among them, prowling in quest of invaders. The enemy tanks crawled upon the same grisly patrol on their own side.

But two miles of the American front had suddenly gone silent. A hundred telephones had ceased to make reports along the line nearest the enemy. As Coffee and Wallis stumbled about the little pill-box, looking for some inkling of the way in which the original occupants of the small strong-point had been wiped out, the second line of observation-posts began to go dead.

Now one, now another abruptly ceased to communicate. Half a dozen were in actual conversation with their sector headquarters, and broke off between words. The wires remained intact. But in fifteen nerve-racking minutes a second hundred posts ceased to make reports and ceased to answer the inquiry-signal. G.H.Q. was demanding explanations in crisp accents that told the matter was being taken very seriously indeed. And then, as the officer in command of the second-line sector headquarters was explaining frenziedly that he was doing all any man could do, he stopped short between two words and thereafter he, also, ceased to communicate.

Front-line sector headquarters seemed inexplicably to have escaped whatever fate had overtaken all its posts, but it could only report that they had apparently gone out of existence without warning. American tanks, prowling in the area that had gone dead, announced that no enemy tanks had been seen. G-81, stumbling on a pill-box no more than ten minutes after it had gone silent, offered to investigate. A member of her crew, in a gas-mask, stepped out of the port doorway. Immediately thereafter G-81's wireless reports stopped coming in.

The situation was clearly shown in the huge tank that had been built to serve as G.H.Q. That tank was seventy feet long, and lay hidden in the mist with a brood of other, smaller tanks clustered near it, from each of which a cable ran to the telephones and instruments of the greater monster. Farther off in the fog, of course, were other tanks, hundreds of them, fighting machines all, silent and motionless now, but infinitely ready to protect the brain of the army.

The G.H.Q. maneuver-board showed the battle as no single observer could ever have seen it. A map lay spread out on a monster board, under a pitiless white light. It was a map of the whole battlefield. Tiny sparks crawled here and there under the map, and there were hundreds of little pins with different-colored heads to mark the position of this thing and that. The crawling sparks were the reported positions of American tanks, made visible as positions of moving trains had been made visible for years on the electric charts of railroads in dispatcher's offices. Where the tiny bulbs glowed under the map, there a tank crawled under the fog. As the tank moved, the first bulb went out and another flashed into light.

The general watched broodingly as the crawling sparks moved from this place to that place, as varicolored lights flashed up and vanished, as a steady hand reached down to shift tiny pins and place new ones. The general moved rarely, and spoke hardly at all. His whole air was that of a man absorbed in a game of chess—a game on which the fate of a nation depended.

He was thus absorbed. The great board, illuminated from above by the glaring bulb, and speckled with little white sparks from below by the tiny bulbs beneath, showed the situation clearly at every instant. The crawling white sparks were his own tanks, each in its present position. Flashing blue sparks noted the last report of enemy tanks. Two staff officers stood behind the general, and each spoke from time to time into a strapped-on telephone transmitter. They were giving routine orders, heading the nearest American patrol-tanks toward the location of the latest reported enemies.

The general reached out his hand suddenly and marked off an area with his fingers. They were long fingers, and slender ones: an artist's fingers.

"Our outposts are dead in this space," he observed meditatively. The use of the word "outposts" dated him many years back as a soldier, back to the old days of open warfare, which had only now come about again. "Penetration of two miles—"

"Tank, sir," said the man of the steady fingers, putting a black pin in position within that area, "let a man out in a gas-mask to examine a pill-box. The tank does not report or reply, sir."

"Gas," said the general, noting the spot. "Their new gas, of course. It must go through masks or sag-paste, or both."

He looked up to one of a row of officers seated opposite him, each man with headphones strapped to his ears and a transmitter before his lips, and each man with a map-pad on his knees, on which from time to time he made notations and shifted pins absorbedly.

"Captain Harvey," said the general, "you are sure that dead spot has not been bombarded with gas-shells?"

"Yes, General. There has been no artillery fire heavy enough to put more than a fraction of those posts out of action, and all that fire, sir, has been accounted for elsewhere."

The officer looked up, saw the general's eyes shift, and bent to his map again, on which he was marking areas from which spotting aircraft reported flashes as of heavy guns beneath the mist.

"Their aircraft have not been dropping bombs, positively?"

A second officer glanced up from his own map.

"Our planes cover all that space, sir, and have for some time."

"They either have a noiseless tank," observed the general meditatively, "or...."

The steady fingers placed a red pin at a certain spot.

"One observation-post, sir, has reopened communication. Two infantrymen, separated from their command, came upon it and found the machine-gun crew dead, with gas-masks adjusted. No tanks or tracks. They are identified, sir, and are now looking for tank tracks or shells."

The general nodded emotionlessly.

"Let me know immediately."

He fell back to the ceaseless study of the board with its crawling sparks and sudden flashes of light. Over at the left, there were four white sparks crawling toward a spot where a blue flash had showed a little while since. A red light glowed suddenly where one of the white sparks crawled. One of the two officers behind the general spoke crisply. Instantly, it seemed, the other three white sparks changed their direction of movement. They swung toward the red flash—the point where a wireless from the tank represented by the first white flash had reported, contact with the enemy.

"Enemy tank destroyed here, sir," said the voice above the steady fingers.

"Wiped out three of our observation posts," murmured the general, "His side knows it. That's an opportunity. Have those posts reoccupied."

"Orders given, sir," said a staff officer from behind. "No reports as yet."

The general's eyes went back to the space two miles wide and two miles deep in which there was only a single observation-post functioning, and that in charge of two strayed infantrymen. The battle in the fog was in a formative stage, now, and the general himself had to watch the whole, because it was by small and trivial indications that the enemy's plans would be disclosed. The dead area was no triviality, however. Half a dozen tanks were crawling through it, reporting monotonously that no sign of the enemy could be found. One of the little sparks representing those tanks abruptly went out.

"Tank here, sir, no longer reports."

The general watched with lack-luster eyes, his mind withdrawn in thought.

"Send four helicopters," he said slowly, "to sweep that space. We'll see what the enemy does."

One of the seated officers opposite him spoke swiftly. Far away a roaring set up and was stilled. The helicopters were taking off.

They would rush across the blanket of fog, their vertical propellers sending blasts of air straight downward. For most of their sweep they would keep a good height, but above the questionable ground they would swoop down to barely above the fog-blanket. There their monstrous screws would blow holes in the fog until the ground below was visible. If any tanks crawled there, in the spaces the helicopters swept clear, they would be visible at once and would be shelled by batteries miles away, batteries invisible under the artificial cloud-bank.

No other noises came through the walls of the monster tank. There was a faint, monotonous murmur of the electric generator. There were the quiet, crisp orders of the officers behind the general, giving the routine commands that kept the fighting a stalemate.

The aircraft officer lifted his head, pressing his headphones tightly against his ears, as if to hear mores clearly.

"The enemy, sir, has sent sixty fighting machines to attack our helicopters. We sent forty single-seaters as escort."

"Let them fight enough," said the general absently, "to cause the enemy to think us desperate for information. Then draw them off."

There was silence again. The steady fingers put pins here and there. An enemy tank destroyed here. An American tank encountered an enemy and ceased to report further. The enemy sent four helicopters in a wide sweep behind the American lines, escorted by fifty fighting planes. They uncovered a squadron of four tanks, which scattered like insects disturbed by the overturning of a stone. Instantly after their disclosure a hundred and fifty guns, four miles away, were pouring shells about the place where they had been seen. Two of the tanks ceased to report.

The general's attention was called to a telephone instrument with its call-light glowing.

"Ah," said the general absently. "They want publicity matter."

The telephone was connected to the rear, and from there to the Capital. A much-worried cabinet waited for news, and arrangements were made and had been used, to broadcast suitably arranged reports from the front, the voice of the commander-in-chief in the field going to every workshop, every gathering-place, and even being bellowed by loud-speakers in the city streets.

The general took the phone. The President of the United States was at the other end of the wire, this time.

"General?"

"Still in a preliminary stage, sir," said the general, without haste. "The enemy is preparing a break-through effort, possibly aimed at our machine-shops and supplies. Of course, if he gets them we will have to retreat. An hour ago he paralyzed our radios, not being aware, I suppose, of our tuned earth-induction wireless sets. I daresay he is puzzled that our communications have not fallen to pieces."

"But what are our chances?" The voice of the President was steady, but it was strained.

"His tanks outnumber ours two to one, of course, sir," said the general calmly. "Unless we can divide his fleet and destroy a part of it, of course we will be crushed in a general combat. But we are naturally trying to make sure that any such action will take place within point-blank range of our artillery, which may help a little. We will cut the fog to secure that help, risking everything, if a general engagement occurs."

There was silence.

The President's voice, when it came, was more strained still.

"Will you speak to the public, General?"

"Three sentences. I have no time for more."

There were little clickings on the line, while the general's eyes returned to the board that was the battlefield in miniature. He indicated a spot with his finger.

"Concentrate our reserve-tanks here," he said meditatively. "Our fighting aircraft here. At once."

The two spots were at nearly opposite ends of the battle field. The chief of staff, checking the general's judgment with the alert suspicion that was the latest addition to his duties, protested sharply.

"But sir, our tanks will have no protection against helicopters!"

"I am quite aware of it," said the general mildly.

He turned to the transmitter. A thin voice had just announced at the other end of the wire, "The commander-in-chief of the army in the field will make a statement."

The general spoke unhurriedly.

"We are in contact with the enemy, have been for some hours. We have lost forty tanks and the enemy, we think, sixty or more. No general engagement has yet taken place, but we think decisive action on the enemy's part will be attempted within two hours. The tanks in the field need now, as always, ammunition, spare tanks, and the special supplies for modern warfare. In particular, we require ever-increasing quantities of fog-gas. I appeal to your patriotism for reinforcements of material and men."

He hung up the receiver and returned to his survey of the board.

"Those three listening-posts," he said abruptly, indicating a place near where an enemy tank had been destroyed. "Have they been reoccupied?"

"Yes, sir. Just reported. The tank they reported rolled over them, destroying the placement. They are digging in."

"Tell me," said the general, "when they cease to report again. They will."

He watched the board again and without lifting his eyes from it, spoke again.

"That listening-post in the dead sector, with the two strayed infantrymen in it. Was it reported?"

"Not yet, sir."

"Tell me immediately it does."

The general leaned back in his chair and deliberately relaxed. He lighted a cigar and puffed at it, his hands quite steady. Other officers, scenting the smoke, glanced up enviously. But the general was the only man who might smoke. The enemy's gases, like the American ones, could go through any gas-mask if in sufficient concentration. The tanks were sealed like so many submarines, and opened their interiors to the outer air only after that air had been thoroughly tested and proven safe. Only the general might use up more than a man's allowance for breathing.

The general gazed about him, letting his mind rest from its intense strain against the greater strain that would come on it in a few minutes. He looked at a tall blond man who was surveying the board intently, moving away, and returning again, his forehead creased in thought.

The general smiled quizzically. That man was the officer appointed to I. I. duty—interpretative intelligence—chosen from a thousand officers because the most exhaustive psychological tests had proven that his brain worked as nearly as possible like that of the enemy commander. His task was to take the place of the enemy commander, to reconstruct from the enemy movements reported and the enemy movements known as nearly as possible the enemy plans.

"Well, Harlin," said the general, "Where will he strike?"

"He's tricky, sir," said Harlin. "That gap in our listening-posts looks, of course, like preparation for a massing of his tanks inside our lines. And it would be logical that he fought off our helicopters to keep them from discovering his tanks massing in that area."

The general nodded.

"Quite true," he admitted. "Quite true."

"But," said Harlin eagerly. "He'd know we could figure that out. And he may have wiped out listening posts to make us think he was planning just so. He may have fought off our helicopters, not to keep them from discovering his tanks in there, but to keep them from discovering that there were no tanks in there!"

"My own idea exactly," said the general meditatively. "But again, it looks so much like a feint that it may be a serious blow. I dare not risk assuming it to be a feint only."

He turned back to the board.

"Have those two strayed infantrymen reported yet?" he asked sharply.

"Not yet, sir."

The general drummed on the table. There were four red flashes glowing at different points of the board—four points where American tanks or groups of tanks were locked in conflict with the enemy. Somewhere off in the enveloping fog that made all the world a gray chaos, lumbering, crawling monsters rammed and battered at each other at infinitely short range. They fought blindly, their guns swinging menacingly and belching lurid flames into the semi-darkness, while from all about them dropped the liquids that meant death to any man who breathed their vapor. Those gases penetrated any gas-mask, and would even strike through the sag-pastes that had made the vesicatory gases of 1918 futile.

With tanks by thousands hidden in the fog, four small combats were kept up, four only. Battles fought with tanks as the main arm are necessarily battles of movement, more nearly akin to cavalry battles than any other unless it be fleet actions. When the main bodies come into contact, the issue is decided quickly. There can be no long drawn-out stalemates such as infantry trenches produced in years past. The fighting that had taken place so far, both under the fog and aloft in the air, was outpost skirmishing only. When the main body of the enemy came into action it would be like a whirlwind, and the battle would be won or lost in a matter of minutes only.

The general paid no attention to those four conflicts, or their possible meaning.

"I want to hear from those two strayed infantrymen," he said quietly, "I must base my orders on what they report. The whole battle, I believe, hinges on what they have to say."

He fell silent, watching the board without the tense preoccupation he had shown before. He knew the moves he had to make in any of three eventualities. He watched the board to make sure he would not have to make those moves before he was ready. His whole air was that of waiting: the commander-in-chief of the army of the United States, waiting to hear what he would be told by two strayed infantrymen, lost in the fog that covered a battlefield.

The fog was neither more dense nor any lighter where Corporal Wallis paused to roll his pre-war cigarette. The tobacco came from the gassed machine-gunner in the pill-box a few yards off. Sergeant Coffee, three yards distant, was a blurred figure. Corporal Wallis put his cigarette into his mouth, struck his match, and puffed delicately.

"Ah!" said Corporal Wallis, and cheered considerably. He thought he saw Sergeant Coffee moving toward him and ungenerously hid his cigarette's glow.

Overhead, a machine-gun suddenly burst into a rattling roar, the sound sweeping above them with incredible speed. Another gun answered it. Abruptly, the whole sky above them was an inferno of such tearing noises and immediately after they began a multitudinous bellowing set up. Airplanes on patrol ordinarily kept their engines muffled, in hopes of locating a tank below them by its noise. But in actual fighting there was too much power to be gained by cutting out the muffler for any minor motive to take effect. A hundred aircraft above the heads of the two strayed infantrymen were fighting madly about five helicopters. Two hundred yards away, one fell to the earth with a crash, and immediately afterward there was a hollow boom. For an instant even the mist was tinged with yellow from the exploded gasoline tank. But the roaring above continued — not mounting, as in a battle between opposing patrols of fighting planes, when each side finds height a decisive advantage, but keeping nearly to the same level, little above the bank of cloud.

Something came down, roaring, and struck the earth no more than fifty yards away. The impact was terrific, but after it there was dead silence while the thunder above kept on.

Sergeant Coffee came leaping to Corporal Wallis' side.

"Helicopters!" he barked. "Huntin' tanks an' pill-boxes! Lay down!"

He flung himself down to the earth.

Wind beat on them suddenly, then an outrageous blast of icy air from above. For an instant the sky lightened. They saw a hole in the mist, saw the little pill-box clearly, saw a huge framework of supporting screws sweeping swiftly overhead with figures in it watching the ground through wind-angle glasses, and machine-gunners firing madly at dancing things in the air. Then it was gone.

"One o' ours," shouted Coffee in Wallis' ear. "They' tryin' to find th' Yellows' tanks!"

The center of the roaring seemed to shift, perhaps to the north. Then a roaring drowned out all the other roarings. This one was lower down and approaching in a rush. Something swooped from the south, a dark blotch in the lighter mist above. It was an airplane flying in the mist, a plane that had dived into the fog as into oblivion. It appeared, was gone—and there was a terrific crash. A shattering roar drowned out even the droning tumult of a hundred aircraft engines. A sheet of flame flashed up, and a thunderous detonation.

"Hit a tree," panted Coffee, scrambling to his feet again. "Suicide club, aimin' for our helicopter."

Corporal Wallis was pointing, his lips drawn back in a snarl.

"Shut up!" he whispered. "I saw a shadow against that flash! Yeller infantryman! Le's get 'im!"

"Y'crazy," said Sergeant Coffee, but he strained his eyes and more especially his ears.

It was Coffee who clutched Corporal Wallis' wrist and pointed. Wallis could see nothing, but he followed as Coffee moved silently through the gray mist. Presently he too, straining his eyes, saw an indistinct movement.

The roaring of motors died away suddenly. The fighting had stopped, a long way off, apparently because the helicopters had been withdrawn.

Except for the booming of artillery a very long distance away, firing unseen at an unseen target, there was no noise at all.

"Aimin' for our pill-box," whispered Coffee.

They saw the dim shape, moving noiselessly, halt. The dim figure seemed to be casting about for something. It went down on hands and knees and crawled forward. The two infantrymen crept after it. It stopped, and turned around. The two dodged to one side in haste. The enemy infantryman crawled off in another direction, the two Americans following him as closely as they dared.

He halted once more, a dim and grotesque figure in the fog. They saw him fumbling in his belt. He threw something, suddenly. There was a little tap as of a fountain pen dropped upon concrete. Then a hissing sound. That was all, but the enemy infantryman waited, as if listening....

The two Americans fell upon him as one individual. They bore him to the earth and Coffee dragged at his gas-mask, good tactics in a battle where every man carries gas-grenades. He gasped and fought desperately, in a seeming frenzy of terror.

They squatted over him, finally, having taken away his automatics, and Coffee worked painstakingly to get off his gas-mask while Wallis went poking about in quest of tobacco.

"Dawggone!" said Coffee. "This mask is intricate."

"He ain't got any pockets," mourned Wallis.

Then they examined him more closely.

"It's a whole suit," explained Coffee. "H-m.... He don't have to bother with sag-paste. He's got him on a land diving-suit."

"S-s-say," gasped the prisoner, his language utterly colloquial in spite of the beady eyes and coarse black hair that marked him racially as of the enemy, "say, don't take off my mask! Don't take off my mask!"

"He talks an' everything," observed Coffee in mild amazement. He inspected the mask again and painstakingly smashed the goggles. "Now, big boy, you take your chance with th' rest of us. What' you doin' around here?"

The prisoner set his teeth, though deathly pale, and did not reply.

"H'm-m...." said Coffee meditatively. "Let's take him in the pill-box an' let Loot'n't Madison tell us what to do with him."

They picked him up.

"No! No! For Gawd's sake, no!" cried the prisoner shrilly. "I just gassed it!"

The two halted. Coffee scratched his nose.

"Reckon he's lyin', Pete?" he asked.

Corporal Wallis shrugged gloomily.

"He ain't got any tobacco," he said morosely. "Let's chuck him in first an' see."

The prisoner wriggled until Coffee put his own automatic in the small of his back.

"How long does that gas last?" he asked, frowning. "Loot'n't Madison wants us to report. There's some fellers in there, all gassed up, but we were in there a while back an' it didn't hurt us. How long does it last?"

"Fur-fifteen minutes, maybe twenty," chattered the prisoner. "Don't put me in there!"

Coffee scratched his nose again and looked at his wrist-watch.

"A'right," he conceded, "we give you twenty minutes. Then we chuck you down inside. That is, if you act real agreeable until then. Got anything to smoke?"

The prisoner agonizedly opened a zipper slip in his costume and brought out tobacco, even tailor-made cigarettes. Coffee pounced on them one second before Wallis. Then he divided them with absorbed and scrupulous fairness.

"Right," said Sergeant Coffee comfortably. He lighted up. "Say, you, if y' want to smoke, here's one o' your pills. Let's see the gas stuff. How' y' use it?"

Wallis had stripped off a heavy belt about the prisoner's waist and it was trailing over his arm. He inspected it now. There were twenty or thirty little sticks in it, each one barely larger than a lead pencil, of dirty gray color, and each one securely nested in a tube of flannel-lined papier-mache.

"These things?" asked Wallis contentedly. He was inhaling deeply with that luxurious enjoyment a tailor-made cigarette can give a man who had been remaking butts into smokes for days past.

"Don't touch 'em," warned the prisoner nervously. "You broke my goggles. You throw 'em, and they light and catch fire, and that scatters the gas."

Coffee touched the prisoner, indicating the ground, and sat down, comfortably smoking one of the prisoner's cigarettes. By his air, he began to approve of his captive.

"Say, you," he said curiously, "you talk English pretty good. How'd you learn it?"

"I was a waiter," the prisoner explained. "New York. Corner Forty-eighth and Sixth."

"My Gawd!" said Coffee. "Me, I used to be a movie operator along there. Forty-ninth. Projection room stuff, you know. Say, you know Heine's place?"

"Sure," said the prisoner. "I used to buy Scotch from that blond feller in the back room. With a benzine label for a prescription?"

Coffee lay back and slapped his knee.

"Ain't it a small world?" he demanded. "Pete, here, he ain't never been in any town bigger than Chicago. Ever in Chicago?"

"Hell," said Wallis, morose yet comfortable with a tailor-made cigarette. "If you guys want to start a extra war, go to knockin' Chicago. That's all."

Coffee looked at his wrist-watch again.

"Got ten minutes yet," he observed. "Say, you must know Pete Hanfry—"

"Sure I know him," said the enemy prisoner, scornfully. "I waited on him. One day, just before us reserves were called back home...."

In the monster tank that was headquarters the general tapped his fingers on his knees. The pale white light flickered a little as it shone on the board where the bright sparks crawled. White sparks were American tanks. Blue flashes were for enemy tanks sighted and reported, usually in the three-second interval between their identification and the annihilation of the observation-post that had reported them. Red glows showed encounters between American and enemy tanks. There were a dozen red glows visible, with from one to a dozen white sparks hovering about them. It seemed as if the whole front line were about to burst into a glare of red, were about to become one long lane of conflicts in impenetrable obscurity, where metal monsters roared and rumbled and clanked one against the other, bellowing and belching flame and ramming each other savagely, while from them dripped the liquids that made their breath mean death. There were nightmarish conflicts in progress under the blanket of fog, unparalleled save perhaps in the undersea battles between submarines in the previous European war.

The chief of staff looked up; his face drawn.

"General," he said harshly, "it looks like a frontal attack all along our line."

The general's cigar had gone out. He was pale, but calm with an iron composure.

"Yes," he conceded. "But you forget that blank spot in our line. We do not know what is happening there."

"I am not forgetting it. But the enemy outnumbers us two to one—"

"I am waiting," said the general, "to hear from those two infantrymen who reported some time ago from a listen-post in the dead area."

The chief of staff pointed to the outline formed by the red glows where tanks were battling.

"Those fights are keeping up too long!" he said sharply. "General, don't you see, they're driving back our line, but they aren't driving it back as fast as if they were throwing their whole weight on it! If they were making a frontal attack there, they'd wipe out the tanks we have facing them; they'd roll right over them! That's a feint! They're concentrating in the dead space—"

"I am waiting," said the general softly, "to hear from those two infantrymen." He looked at the board again and said quietly, "Have the call-signal sent them. They may answer."

He struck a match to relight his dead cigar. His fingers barely quivered as they held the match. It might have been excitement—but it might have been foreboding, too.

"By the way," he said, holding the match clear, "have our machine-shops and supply-tanks ready to move. Every plane is, of course, ready to take the air on signal. But get the aircraft ground personnel in their traveling tanks immediately."

Voices began to murmur orders as the general puffed. He watched the board steadily.

"Let me know if anything is heard from these infantrymen...."

There was a definite air of strain within the tank that was headquarters. It was a sort of tensity that seemed to emanate from the general himself.

Where Coffee and Wallis and the prisoner squatted on the ground, however, there was no sign of strain at all. There was a steady gabble of voices.

"What kinda rations they give you?" asked Coffee interestedly.

The enemy prisoner listed them, with profane side-comments.

"Hell," said Wallis gloomily. "Y'ought to see what we get! Las' week they fed us worse'n dogs. An' th' canteen stuff—"

"Your tank men, they get treated fancy?" asked the prisoner.

Coffee made a reply consisting almost exclusively of high powered expletives.

"—and the infantry gets it in the neck every time," he finished savagely. "We do the work—"

Guns began to boom, far away. Wallis cocked his ears.

"Tanks gettin' together," he judged, gloomily. "If they'd all blow each other to hell an' let us infantry fight this battle—"

"Damn the tanks!" said the enemy prisoner viciously. "Look here, you fellers. Look at me. They sent a battalion of us out, in two waves. We hike along by compass through the fog, supposed to be five paces apart. We come on a pill-box or listenin' post, we gas it an' go on. We try not to make a noise. We try not to get seen before we use our gas. We go on, deep in your lines as we can. We hear one of your tanks, we dodge it if we can, so we don't get seen at all. O'course we give it a dose of gas in passing, just in case. But we don't get any orders about how far to go or how to come back. We ask for recognition signals for our own tanks, an' they grin an' say we won't see none of our tanks till the battle's over. They say 'Re-form an' march back when the fog is out.' Ain't that pretty for you?"

"You second wave?" asked Coffee, with interest.

The prisoner nodded.

"Mopping up," he said bitterly, "what the first wave left. No fun in that! We go along gassin' dead men, an' all the time your tanks is ravin' around to find out what's happenin' to their listenin'-posts. They run into us—"

Coffee nodded sympathetically.

"The infantry always gets the dirty end of the stick," said Wallis morosely.

Somewhere, something blew up with a violent explosion. The noise of battle in the distance became heavier and heavier.

"Goin' it strong," said the prisoner, listening.

"Yeh," said Coffee. He looked at his wrist-watch. "Say, that twenty minutes is up. You go down in there first, big boy."

They stood beside the little pill-box. The prisoner's knees shook.

"Say, fellers," he said pleadingly, "they told us that stuff would scatter in twenty minutes, but you busted my mask. Yours ain't any good against this gas. I'll have to go down in there if you fellers make me, but—"

Coffee lighted another of the prisoner's tailor-made cigarettes.

"Give you five minutes more," he said graciously. "I don't suppose it'll ruin the war."

They sat down relievedly again, while the fog-gas made all the earth invisible behind a pall of grayness, a grayness from which the noises of battle came.

In the tank that was headquarters, the air of strain was pronounced. The maneuver-board showed the situation as close to desperation, now. The reserve-tank positions had been switched on the board, dim orange glows, massed in curiously precise blocks. And little squares of green showed there that the supply and machine-shop tanks were massed. They were moving slowly across the maneuver-board. But the principal change lay in the front-line indications.

The red glows that showed where tank battles were in progress formed an irregularly curved line, now. There were twenty or more such isolated battles in progress, varying from single combats between single tanks to greater conflicts where twenty to thirty tanks to a side were engaged. And the positions of those conflicts were changing constantly, and invariably the American tanks were being pushed back.

The two staff officers behind the general were nearly silent. There were few sparks crawling within the American lines now. Nearly every one had been diverted into the front-line battles. The two men watched the board with feverish intensity, watching the red glows moving back, and back....

The chief of staff was shaking like a leaf, watching the American line stretched, and stretched....

The general looked at him with a twisted smile.

"I know my opponent," he said suddenly. "I had lunch with him once in Vienna. We were attending a disarmament conference." He seemed to be amused at the ironic statement. "We talked war and battles, of course. And he showed me, drawing on the tablecloth, the tactical scheme that should have been used at Cambrai, back in 1917. It was a singularly perfect plan. It was a beautiful one."

"General," burst out one of the two staff officers behind him. "I need twenty tanks from the reserves."

"Take them," said the general. He went on, addressing his chief of staff. "It was an utterly flawless plan. I talked to other men. We were all pretty busy estimating each other there, we soldiers. We discussed each other with some freedom, I may say. And I formed the opinion that the man who is in command of the enemy is an artist: a soldier with the spirit of an amateur. He's a very skilful fencer, by the way. Doesn't that suggest anything?"

The chief of staff had his eyes glued to the board.

"That is a feint, sir. A strong feint, yes, but he has his force concentrated in the dead area."

"You are not listening, sir," said the general, reprovingly. "I am saying that my opponent is an artist, an amateur, the sort of person who delights in the delicate work of fencing. I, sir, would thank God for the chance to defeat my enemy. He has twice my force, but he will not be content merely to defeat me. He will want to defeat me by a plan of consummate artistry, which will arouse admiration among soldiers for years to come."

"But General, every minute, every second—"

"We are losing men, of whom we have plenty, and tanks, of which we have not enough. True, very true," conceded the general. "But I am waiting to hear from two strayed infantrymen. When they report, I will speak to them myself."

"But, sir," cried the chief of staff, withheld only by the iron habit of discipline from violent action and the taking over of command himself, "they may be dead! You can't risk this battle waiting for them! You can't risk it, sir! You can't!"

"They are not dead," said the general coolly. "They cannot be dead. Sometimes, sir, we must obey the motto on our coins. Our country needs this battle to be won. We have got to win it, sir! And the only way to win it—"

The signal-light at his telephone glowed. The general snatched it up, his hands quivering. But his voice, was steady and deliberate as he spoke.

"Hello, Sergeant—Sergeant Coffee, is it?... Very well, Sergeant. Tell me what you've found out.... Your prisoner objects to his rations, eh? Very well, go on.... How did he gas our listening-posts?... He did, eh? He got turned around and you caught him wandering about?... Oh, he was second wave! They weren't taking any chances on any of our listening-posts reporting their tanks, eh?... Say that again, Sergeant Coffee!" The general's tone had changed indescribably. "Your prisoner has no recognition signals for his own tanks? They told him he wouldn't see any of them until the battle was over?... Thank you, Sergeant. One of our tanks will stop for you. This is the commanding general speaking."

He rang off, his eyes blazing. Relaxation was gone. He was a dynamo, snapping orders.

"Supply tanks, machine-shop tanks, ground forces of the air service, concentrate here!" His finger rested on a spot in the middle of the dead area. "Reserve tanks take position behind them. Draw off every tank we've got—take 'em out of action!—and mass them in front, on a line with our former first line of outposts. Every airplane and helicopter take the air and engage in general combat with the enemy, wherever the enemy may be found and in whatever force. And our tanks move straight through here!"

Orders were snapping into telephone transmitters. The commands had been relayed before their import was fully realized. Then there was a gasp.

"General!" cried the chief of staff. "If the enemy is massed there, he'll destroy our forces in detail as they take position!"

"He isn't massed there," said the general, his eyes blazing. "The infantrymen who were gassing our listening-posts were given no recognition signals for their tanks. Sergeant Coffee's prisoner has his gas-mask broken and is in deadly fear. The enemy commander is foolish in many ways, perhaps, but not foolish enough to break down morale by refusing recognition signals to his own men who will need them. And look at the beautiful plan he's got."

He sketched half a dozen lines with his fingers, moving them in lightning gestures as his orders took effect.

"His main force is here, behind those skirmishes that look like a feint. As fast as we reinforce our skirmishing-line, he reinforces his—just enough to drive our tanks back slowly. It looks like a strong feint, but it's a trap! This dead space is empty. He thinks we are concentrating to face it. When he is sure of it—his helicopters will sweep across any minute, now, to see—he'll throw his whole force on our front line. It'll crumple up. His whole fighting force will smash through to take us, facing the dead space, in the rear! With twice our numbers, he'll drive us before him."

"But general! You're ordering a concentration there! You're falling in with his plans!"

The general laughed.

"I had lunch with the general in command over there, once upon a time. He is an artist. He won't be content with a defeat like that! He'll want to make his battle a masterpiece, a work of art! There's just one touch he can add. He has to have reserves to protect his supply-tanks and machine-shops. They're fixed. The ideal touch, the perfect tactical fillip, will be—Here! Look. He expects to smash in our rear, here. The heaviest blow will fall here. He will swing around our right wing, drive us out of the dead area into his own lines—and drive us on his reserves! Do you see it? He'll use every tank he's got in one beautiful final blow. We'll be outwitted, out-numbered, out-flanked and finally caught between his main body and his reserves and pounded to bits. It is a perfect, a masterly bit of work!"

He watched the board, hawklike.

"We'll concentrate, but our machine-shops and supplies will concentrate with us. Before he has time to take us in rear we'll drive ahead, in just the line he plans for us! We don't wait to be driven into his reserves. We roll into them and over them! We smash his supplies! We destroy his shops! And then we can advance along his line of communication and destroy it, our own depots being blown up—give the orders when necessary—and leaving him stranded with motor-driven tanks, motorized artillery, and nothing to run his motors with! He'll be marooned beyond help in the middle of our country, and we will have him at our mercy when his tanks run out of fuel. As a matter of fact, I shall expect him to surrender in three days."

The little blocks of green and yellow that had showed the position of the reserve and supply-tanks, changed abruptly to white, and began to crawl across the maneuver-board. Other little white sparks turned about. Every white spark upon the maneuver-board suddenly took to itself a new direction.

"Disconnect cables," said the general, crisply. "We move with our tanks, in the lead!"

The monotonous humming of the electric generator was drowned out in a thunderous uproar that was muffled as an air-tight door was shut abruptly. Fifteen seconds later there was a violent lurch, and the colossal tank was on the move in the midst of a crawling, thundering horde of metal monsters whose lumbering progress shook the earth.

Sergeant Coffee, still blinking his amazement, absent-mindedly lighted the last of his share of the cigarettes looted from the prisoner.

"The big guy himself!" he said, still stunned. "My Gawd! The big guy himself!"

A distant thunder began, a deep-toned rumbling that seemed to come from the rear. It came nearer and grew louder. A peculiar quivering seemed to set up in the earth. The noise was tanks moving through the fog, not one tank or two tanks, or twenty tanks, but all the tanks in creation rumbling and lurching at their topmost speed in serried array.

Corporal Wallis heard, and turned pale. The prisoner heard, and his knees caved in.

"Hell," said Corporal Wallis dispairingly. "They can't see us, an' they couldn't dodge us if they did!"

The prisoner wailed, and slumped to the floor.

Coffee picked him up by the collar and jerked him out of the pill-box.

"C'mon Pete," he ordered briefly. "They ain't givin' us a infantryman's chance, but maybe we can do some dodgin'!"

Then the roar of engines, of metal treads crushing upon earth and clinking upon their joints, drowned out all possible other sounds. Before the three men beside the pill-box could have moved a muscle, monster shapes loomed up, rushing, rolling, lurching, squeaking. They thundered past, and the hot fumes of their exhausts enveloped the trio.

Coffee growled and put himself in a position of defiance, his feet braced against the concrete of the pill-box dome. His expression was snarling and angry but, surreptitiously, he crossed himself. He heard the fellows of the two tanks that had roared by him, thundering along in alignment to right and left. A twenty-yard space, and a second row of the monsters came hurtling on, gun muzzles gaping, gas-tubes elevated, spitting smoke from their exhausts that was even thicker than the fog. A third row, a fourth, a fifth....

The universe was a monster uproar. One could not think in this volume of sound. It seemed that there was fighting overhead. Crackling noises came feebly through the reverberating uproar that was the army of the United States in full charge. Something came whirling down through the overhanging mist and exploded in a lurid flare that for a second or two cast the grotesque shadows of a row of tanks clearly before the trio of shaken infantrymen.

Still the tanks came on and roared past. Twenty tanks, twenty-one ... twenty-two.... Coffee lost count, dazed and almost stunned by the sheer noise. It rose from the earth and seemed to be echoed back from the topmost limit of the skies. It was a colossal din, an incredible uproar, a sustained thunder that beat at the eardrums like the reiterated concussions of a thousand guns that fired without ceasing. There was no intermission, no cessation of the tumult. Row after row after row of the monsters roared by, beaked and armed, going greedily with hungry guns into battle.

And then, for a space of seconds, no tanks passed. Through the pandemonium of their going, however, the sound of firing somehow seemed to creep. It was gunfire of incredible intensity, and it came from the direction in which the front-rank tanks were heading.

"Forty-eight, forty-nine, forty-ten, forty-'leven," muttered Coffee dazedly, his senses beaten down almost to unconsciousness by the ordeal of sound. "Gawd! The whole army went by!"

The roaring of the fighting-tanks was less, but it was still a monstrous din. Through it, however, came now a series of concussions that were so close together that they were inseparable, and so violent that they were like slaps upon the chest.

Then came other noises, louder only because nearer. These were different noises, too, from those the fighting-tanks had made. Lighter noises. The curious, misshapen service tanks began to rush by, of all sizes and all shapes. Fuel-carrier tanks. Machine-shop tanks, huge ones, these. Commissary tanks....

Something enormous and glistening stopped short. A door opened. A voice roared an order. The three men, beaten and whipped by noise, stared dumbly.

"Sergeant Coffee!" roared the voice. "Bring your men! Quick!"

Coffee dragged himself back to a semblance of life. Corporal Wallis moved forward, sagging. The two of them loaded their prisoner into the door and tumbled in. They were instantly sent into a heap as the tank took up its progress again with a sudden sharp leap.

"Good man," grinned a sooty-faced officer, clinging to a handhold. "The general sent special orders you were to be picked up. Said you'd won the battle. It isn't finished yet, but when the general says that—"

"Battle?" said Coffee dully. "This ain't my battle. It's a parade of a lot of damn tanks!"

There was a howl of joy from somewhere above. Discipline in the machine-shop tanks was strict enough, but vastly different in kind from the formality of the fighting-machines.

"Contact!" roared the voice again. "General wireless is going again! Our fellows have rolled over their reserves and are smashing their machine-shops and supplies!"

Yells reverberated deafeningly inside the steel walls, already filled with tumult from the running motors and rumbling treads.

"Smashed 'em up!" shrieked the voice above, insane with joy. "Smashed 'em! Smashed 'em! Smashed 'em! We've wiped out their whole reserve and—" A series of detonations came through even the steel shell of the lurching tank. Detonations so violent, so monstrous, that even through the springs and treads of the tank the earth-concussion could be felt. "There goes their ammunition! We set off all their dumps!"

There was sheer pandemonium inside the service-tank, speeding behind the fighting force with only a thin skin of reserve-tanks between it and a panic-stricken, mechanically pursuing enemy.

"Yell, you birds!" screamed the voice. "The general says we've won the battle! Thanks to the fighting force! We're to go on and wipe out the enemy line of communications, letting him chase us till his gas gives out! Then we come back and pound him to bits! Our tanks have wiped him out!"

Coffee managed to find something to hold on to. He struggled to his feet. Corporal Wallis, recovering from the certainty of death and the torture of sound, was being very sea-sick from the tank's motion. The prisoner moved away from him on the steel floor. He looked gloomily up at Coffee.

"Listen to 'em," said Coffee bitterly. "Tanks! Tanks! Tanks! Hell! If they'd given us infantry a chance—"

"You said it," said the prisoner savagely. "This is a hell of a way to fight a war."

Corporal Wallis turned a greenish face to them.

"The infantry always gets the dirty end of the stick," he gasped. "Now they—now they' makin' infantry ride in tanks! Hell!"

Invisible Death

By Anthony Pelcher

On Lees' quick and clever action depended the life of "Old Perk" Ferguson, the millionaire manufacturer threatened by the uncanny, invisible killer.

The inquest into the mysterious death of Darius Darrow, savant, inventor, recluse and eccentric, resembled a scientific convention. Men and women of high scientific attainment, and, in some instances, world fame, attended to hear first hand the strange, uncanny, unbelievable circumstances as hinted by the newspapers.

Mrs. Susan Darrow, the widow, was the paramount witness. She appeared a quaint figure as she took the stand. Tearful, yet alert, this little woman betrayed the intelligence that had made her one of the world's foremost chemists. She gave her age as fifty-eight, but if it had not been for her snowy hair she would have looked much younger. She was small but not frail, and had expressive blue eyes. She had a firm little nose and chin, and was garbed in black silk garments of a fashion evidently dating back a decade.

Although not modern in dress, her answers to questions regarding scientific and business affairs involved in the mysterious case, proved she was thoroughly abreast of the times in all other particulars.

"You believe your husband was murdered?" bluntly asked the examiner at one stage.

"That is my opinion," she said, then added: "It might have been some scientific accident, the nature of which I cannot fathom. We were confidential in all matters except my husband's work. He reserved the right to be secretive about the scientific problems on which he was working."

"Can you throw any light on a motive for such a crime?"

"The motive seems self-evident. He was working on an invention that he said would do away with war and would make the owner of the device a practical world dictator, should he choose to exercise such power. The device was completed. The murderer killed him to secure his device. That all seems plain enough."

"Was anything else of value taken?"

"We had nothing else of value about the place. I was never given to jewelry. The furnishings and equipment were undisturbed. It is quite evident, I think, that the thief was no ordinary petty burglar."

The attorney interposed: "I believe we had better let Mrs. Darrow tell this story from the beginning in her own way. There are only two really important witnesses. Whatever she can remember to recite might be of value to the authorities. Now, Mrs. Darrow, how long had you lived at Brooknook? Begin there and just let your story unfold. Try to control your nerves and emotions."

"I am not emotional. I am not nervous," said the quaint little woman, bravely. "My heart hurts, that is all.

"The place was named by my father. We inherited it at his death, thirty years ago, and moved in. My two children were born and died there. At first we kept the servants and maintained all of the thirty-two rooms. But after the children were gone, we both gave ourselves over to study and we began to close one room after another, releasing the servants one by one."

"How many rooms do you occupy now?"

"We lived in three, a living-room, kitchen and bedroom. The two big parlors were turned into a laboratory. We both worked there. It was there my husband met his death at his work. Sometimes we worked together, sometimes independently. I did all my own housework, except the laundry, which I sent out. We had no visitors. We lived for each other and our work."

"Tell us about the rooms that were not occupied."

"We left them just as they always had been. I have not been in any of these rooms for twenty years. Once I looked into the little girl's room—my daughter's room. It was dusty and cobwebby, but undisturbed by human hand. My husband peered in over my shoulder. I closed the door. We turned away in each other's arms."

Here the little old woman fell to weeping softly into her lace handkerchief. Minutes lapsed as the court waited, respecting her grief.

"Were these rooms locked?" asked the attorney finally.

"No," said the widow, recovering, as she dabbed at her eyes. "We feared no one. All the rooms were closed, but not locked. The outside doors were seldom locked. We lived in our own world. For appearance sake we kept up the grounds. Peck, the gardener, kept the grounds, as you know. He called in outside help when necessary. This was his affair. We never bothered him. He lived probably a half mile up the road. The first of each month he would come for his pay. He was practically our only visitor.

"When it was necessary to see our attorney or other connections, Peck would drive us. At first he used to drive our horses. Ten years ago we pastured the horses for life and bought the small car. We seldom went out. We have no close friends and no relatives nearer than the Pacific coast. They are distant cousins. You see, we were rather alone in the world since the children went away—we never spoke of them as being dead."

Again the court was hushed. The coroner and the attorney took occasion to blow their noses rather violently.

"On May 27th, the day your husband died, what happened, as you re-remember it?" asked the attorney.

"We arose and had breakfast as usual. I was puttering about the rooms. My husband kissed me and started for the laboratory. I was in the kitchen. It was about ten o'clock when I finished in the kitchen and went into the living room which adjoins the laboratory. I had been rather fretted, something unusual for me. It seemed I dimly sensed the presence of someone near me, someone I did not know, an outsider. I thought it was foolish of me and buckled up.

"But when I went into the living room, it seemed as if some invisible presence were following me. I could hear the low hum of my husband's device. The door of the laboratory was open. He called to me and said:

"'Sue dear, it seems strange, but I made two models of this set and now I can find only one. You could not have misplaced the other by any chance, could you?'

"I assured him I knew nothing of it and he said, 'Hum-m, that's funny.' Then he went back into the library and closed the door. The humming continued. I was more annoyed than ever, but I did not want to bother my husband. Then a queer thing happened. I saw the door of the laboratory open and close, but I did not see anyone. The next instant, I heard my husband's outcry. It was more a groan than a scream.

"I rushed into the laboratory. My husband was lying by his slate-topped table. The device, I noticed, was gone. It was no bigger than a coffee-mill, I thought, as I bent over my husband. Strange how such a thought could have crowded in at such a time.

"My husband's head was bleeding. It was cut, a long gash over the ear, just below the bald spot. It must have been a frightful blow. I looked in his eyes. My nurse's and pharmaceutical course gave me knowledge which sent a chill to my heart. He was dead. I must have fainted.

"When I recovered I ran for Peck. I found him near the house, coming my way and holding his right eye.

"'Something struck me,' he said. Then, seeing me so pale, he said, 'My God! Mrs. Darrow, what has happened?'

"'Run for the doctor,' I said. When the doctor came he called the police and coroner. They told me not to disturb the body. Later they took it away, and the gardener told me—"

"Never mind what Peck told you," interrupted the attorney. "We will let him tell it. Is that all you can tell us about the death itself?"

But the widow was weeping now, so violently that the court ordered her excused.

The gardener was called and took the stand displaying a big, black eye, which offered comedy relief to a pathetic situation.

"On the main road to the east," he began after preliminary questioning, "was a small car which had been parked there all morning. I noticed it because it had no license plates. It was visible from the inside of the grounds, but was hidden from the road by a hedge. It made me wonder because it was just inside our grounds.

"I had some very special red flags which I planted as a border back of pink geraniums. They were doing fine. I got them from the Fabrish seed house. There are no plants like Fabrish's—I wouldn't give a snap of my finger for all the other—"

"Just a minute," interrupted the attorney. He told the gardener to never mind the geraniums and flags, but to tell just what happened.

"Well, I was bending over the border bed when I heard sounds like someone running along the gravel path towards me. I heard a humming like a bumble bee and I jumped to my feet. Just then something hit me in the eye and knocked me down. Yes sir, knocked me plumb down, and—"

"Then what happened? Never mind the asides, the extras—tell us just the simple facts," instructed the attorney.

"Well, you won't believe it, but I heard the footsteps leave the road. The geraniums were badly trampled. I looked at the parked automobile and could hear the hum coming from there.

"The machine started and turned into the road—"

"Did you notice anyone at the wheel?"

"That's what you're not going to believe. There wasn't anybody in that auto at all. I didn't see anyone at any time. The auto started itself, and what is more, that auto only went about a hundred yards when it disappeared altogether—like that—like a flash."

"Did it turn off the road?"

"I didn't turn anywhere. It was in the middle of the road. It just disappeared right in the middle of the road. It started without a driver, it turned north without a driver, and went on by itself for about a hundred yards. Then it vanished in the middle of the road. Just dropped out of sight."

The court-room was hushed. The audience and court attaches were awe stricken and looked their incredulity.

"Do you mean to tell us that auto drove itself?" asked the court sternly.

The witness was completely confused. The attorney came to his rescue, looked at the court, and said:

"He has told that same story a hundred times, and he will stick to it. It seems impossible, but has not Mrs. Darrow told us she heard this humming and saw nothing? With the purely perfunctory recitals of the doctor and the constabulary this court and the jury have heard all there is to hear. We have no more witnesses. That is all there is.

"The jury will have to decide from the evidence whether this case is accident or murder. The doctor and two experts have reported that the wound appeared to have been made by some blunt instrument, swung powerfully. The skull under the wound and back of the ear was simply crushed. Death was instantaneous. It all happened in broad daylight."

After an hour's deliberation the jury decided the savant came to his death in his laboratory from a blow on the skull received in some manner unknown.

The crowd filed out, spiritedly discussing the unusual crime. In the crowd was Perkins Ferguson, known as "Old Perk," head of the Schefert Engineering Corporation, who paid royalty on some of the Darrow patents. With him was Damon Farnsworth, his first vice-president.

"Well, what do you think of it?" asked Farnsworth, biting into a black cigar.

"Damned weird, isn't it?" replied "Old Perk." "I have my own theory, however," he added, "but I am going to know a whole lot more about this case before I venture it." The pair climbed into Ferguson's car discussing the Darrow death case with furrowed brows.

What might be termed an extraordinary meeting of the directors of the Schefert Engineering Corporation, was held a few days later in a big building in the financial district.

The rich furnishings of the directors' room indicated, better than Bradstreet's, the great wealth of the corporation. Uniformed pages stood at attention at each end of the long, mahogany table at which were seated the fourteen directors of the company. All were men of wealth, standing and engineering knowledge. The departed Darrow often had been summoned to such meetings, and at this one there was a hush because of his recent demise.

After a batch of preliminary business had been transacted, Ferguson arose and cleared his throat. The directors leaned forward in their chairs expectantly. The page boys lost their mechanical attitude for the instant and fairly craned their necks around the bulks of the forms in front of them.

"The Darrow case has taken a sudden and sinister turn," said the president. "I have a letter. I will read it:

> "Old Perk: Get wise to yourself. We are in a position to destroy you and all the pot-bellies in the Wall Street crowd. If you want to die of old age, remember what happened to Darrow and begin declaring us in on Wall Street dividends. If you do not you will follow Darrow in the same way.
>
> "Our first demand is for $100,000. Leave this amount in hundreds and fifties in the rubbish can at the corner of 50th Street and Broadway at 10 A. M. next Thursday. If you fail we will break your damned neck. Bring the police with you if you like.
>
> Invisible Death.

Ferguson passed the letter around for inspection. It was painstakingly printed, evidently from the type in a rubber stamp set such as is sold in toy stores.

"I have decided," said Perkins at length, "to give this case to Walter Lees. He has never failed us in mechanical, chemical, or any form of scientific problem. I hope he will not fail in this. He will work independently of the police, who have requested that we keep the appointment at 50th Street and Broadway at the hour named. We will deposit a roll of newspapers, around which has been wrapped a fifty dollar bill and then we will stand by while the awaiting detectives do their duty."

"You do not think anyone is going to call for any supposed package of money at one of the most congested corners in the world in broad daylight?" asked a director at the end of the table.

"Why not?" asked Ferguson. "A seedy individual could pick a package from a rubbish bin at that corner without attracting the least attention."

"I guess you're right," agreed the doubting one.

"I know I'm right," said the president. And he usually was.

"I have already arranged to have Lees instructed in his work," Ferguson volunteered as a pause came in the buzz of conversation about the table. "Lees is young, but he is capable." There was general discussion of the strange case of Darius Darrow; the room filled with the blue haze of many cigars.

Suddenly a low, humming sound was heard in the room.

Papers on the directors' table were bunched as if by unseen hands, and thrown to the ceiling, from which they descended like flakes of snow and scattered about the room.

A book of minutes was torn from the hands of a secretary. It was raised and brought down on vice-president Farnsworth's head. A chair was pulled out from under another director and he was deposited in an undignified heap on the floor.

Another director acted as though he had been tripped, and he fell on top of Farnsworth. Two big vases crashed to the floor in bits. Other decorative objects were scattered about.

The directors who had been hurtled to the floor stood up with expressions of comical surprise on their features. Their chairs catapulted into a far corner of the room, one after the other.

Startled expressions resounded from the group.

A small bookcase fell on its front with a crash of glass. Ferguson's cane jumped in the air and crashed a window pane.

The humming ceased suddenly.

The room was a wreck. The assembled men stood aghast. They were simply nonplussed. Finally they phoned for the police.

After hearing the strange recital from so many highly reputable witnesses, a detective sergeant, who had responded to the call with others, reported to headquarters.

A uniformed police guard was sent to the place with instructions to remain on duty until relieved.

Ferguson sent for Walter Lees, the young engineer of whom he had spoken to the directorate. Assigned to the task of unraveling the Darrow death mystery, Lees ran true to form by getting busy at once. This was at midnight of the day of the surprising directors' meeting. Lees owned a big car; he piled into it and started for the scene of the crime.

Daybreak found him examining every inch of the road around the Darrow estate. Then he searched the hedge along the east road, where the phantom auto had disappeared after the crime. The brush along the opposite side of the thoroughfare was also gone over.

Passing autos had stopped to ask the meaning of his flashlight. Lees explained he had lost a pocketbook. It was as good an excuse as any and served to keep him from drawing a crowd. He found nothing to reward his long and painstaking efforts.

At seven A. M. he decided to interview the Darrow widow, and found her already up and about her kitchen, weeping softly as she worked.

She bade him be seated in the living room.

"No, I am not afraid to stay here alone," she said in reply to Lees' first question. "Whoever killed my husband did so to get possession of his second model. They had already stolen the first. I have thought since that they were afraid that the finding of the second model after his death would aid in their detection. For some reason they had to have both models."

She agreed to tell all she knew of the case. Lees listened to the long recital as already recorded at the coroner's inquest. By adroit questioning Lees gained just one new fact. Mrs. Darrow remembered that she had called her husband, just before he retired to his laboratory, to fix a towel hanger in the kitchen. "He found the pivot needed oiling," explained the widow. "That was all. He oiled it and went into the laboratory."

The idea of one of the world's greatest mechanical engineers stopping his work to oil a towel hanger caused Lees to smile, but Mrs. Darrow did not smile.

"My husband was a genius at repairing about the house," she said, in all seriousness.

"I can imagine so," agreed Lees.

The conversation ceased. Lees sat for a few minutes with his head in his hands, thinking deeply. Finally he said:

"I am convinced that someone who was well aware of your husband's habits committed this crime. Do you believe, positively, that the gardener is above suspicion?"

"Oh, it couldn't have been Peck," insisted Mrs. Darrow. "I had seen him down near the gate from the window. He was too far from the house, and besides, he was devoted to us both."

"Then it was somebody from the neighborhood," said Lees.

"Maybe so," replied Mrs. Darrow, noncommittally.

"Who lives in the next house south?"

"That is towards the city," mused the widow. "There are no houses south on either side of the road for a little further than a mile, when you reach the town limits of Farsdale. The town line is about half-way between, and marks the southern end of this estate."

"Who lives in the first house to the north?"

"That is the cottage of Peck, the gardener."

"How near is the next house?"

"That was the parcel my father sold. It is about three acres, and in the center, or about the center, is the house built by Adolph Jouret, who bought the land. He lives there with his daughter. They built a magnificent place. The brook that traverses our grounds rises at a spring back of his house. Save for two West Indian servants, they are alone. The servants live in Farsdale and motor back and forth."

"What do you know of this—what's his name?" queried Lees, who had assumed the role of examiner.

"Jouret? Very little. He is some sort of a circus man or showman, or was before he retired. He once had wealth, but my husband, some weeks ago, said that because of ill-advised investments he was not so well rated as formerly. I had the feeling that he might be forced to give up the place. I just felt that. I never heard it. I am so sorry because of the daughter. She is a beautiful girl, and seemed kindly, the one time I saw her. She was about twelve then. I do not like to say it, but she seemed a little dazed or slow witted, but really beautiful." Mrs. Darrow fell to smoothing out the folds in her house apron as Lees asked:

"When was the only time you saw her?"

"Ten years ago, about. Just after my father's death. They called on us. We did not care to continue the friendship, as Jouret seemed a little flamboyant—his circus nature, I suppose. Anyway, we were quiet folks, and there was no need of close association with neighbors.

"I remember," continued the widow, after a pause, "that Jouret, when he heard my husband was a scientist, simulated an interest in science. He did have a smattering knowledge of science, but he was plainly affected, so we decided to just let him drop. No ill-feeling. We just—well, we were not interested."

"You do not approve of circus people?"

"It is not that. Any honest work is honorable. It seems commendable to furnish amusement for the public. I know little about people of his profession but I am sure they are perfectly all right. It was Jouret, personally. He seemed noisy and insincere. The girl was nice. I loved her."

"That is all you know of the Jourets?"

"That is all."

"Mrs. Darrow, I wish to go through this house from attic to basement. Have you any objections?"

"None whatever. Make yourself free, but do not attach any significance to what appears to be a secret passageway and cave. My father was a biological chemist. He used to experiment much with small animals. He had a cave where he stored chemicals, and I believe you will find old chemicals stored down there now. I disturbed nothing."

The widow forced a smile to her lips. "Will you excuse me?" she concluded. "I am trying to carry on."

Lees, carrying a flashlight, began a systematic search of the premises. He made his way up a winding staircase, through dust and cobwebs to the attic. He found the top story filled with trunks and bits of furniture of a previous generation. All was in order, but dust-covered and cobwebby.

"Someone has been here before me," he said to himself, brushing a mist of cobwebs from his coat sleeves. "There is a path brushed through the spiderwebs." Turning his flashlight on the floor, he exclaimed:

"And here are footprints in the dust. Well I'll be—!"

Then, after some study, he mused:

"Of course there has been someone here. The killer of Darrow probably has been here to see what he could see. It was no great task. The doors were never locked. The footprints are of no value except to give me the size of his shoes."

He measured the footprints carefully. Then he went downstairs and phoned the measurements to a local shoe dealer, asking him to give him the trade size of shoes which would make such prints.

"They are number nines," decided the shoe dealer.

Lees then returned to resume his search in the rooms and corridors.

"Wonder if Jouret wears nines," he questioned himself. "But what if he does? I couldn't convict him on that score. However, it might help."

Then he fell to searching through the old trunks. He found old photographs, articles of apparel, knicknacks—grandmother's and grandfather's belongings all of them, and some children's clothes of the days when little boys wore ruffles about their necks and little girls' pantalettes reached to their ankles.

Carefully each article was replaced. He made his way down to the third and then the second floor. Through cobwebby corridors and bedchambers he searched, but found nothing further to aid his case.

In the unused rooms on the first floor he found an old spinning-wheel, candle moulds and utensils used in cooking in the days when housewives cooked over an open fire.

He did not find the "secret" passageway until Mrs. Darrow came to his aid. Leading from the basement was a coal chute. This shoot was formed in a triangle with the point under a trap. It was man-high at the cellar opening and its floor was a slide for fuel. It had been in use, evidently, quite recently.

At the cellar wall of this chute, Mrs. Darrow pressed what appeared to be a knot in the old timber and pushed open a door.

A dank odor issued forth as the door was opened. Lees entered the passage and Mrs. Darrow returned upstairs.

Following the underground passageway, Lees came onto a cave about 14 by 14 feet in size with a ceiling and walls of arched brick. It had evidently been built before the days of cement construction.

A long bench and shelves with carboys and jars of chemicals were the only furnishings. Lees sounded all the walls, but found nothing further to interest him.

Lees returned to town at the urgent call of "Old Perk," who had arranged with great care to keep the appointment at 50th street and Broadway, where the decoy package was to be left. He had snipers in nearby windows. He had detectives, dressed in the gay garb of the habitues of the neighborhood, patrolling the corner, and he and his own guard parked an automobile, against all traffic rule, at the curb near the rubbish can.

An office boy sauntered up to the rubbish can, threw in the decoy package, and sauntered away.

A second later there was a low humming sound. The decoy package fairly jumped out of the rubbish can and disappeared in thin air.

The humming sound seemed to round the corner into 50th Street. Detectives followed on the jump. The humming approached an auto at the curb and the auto's self starter began to function. As the police stood near by, enough to have jumped into the auto, the whole machine, a big touring car, actually disappeared before their eyes.

Consternation is a mild word when used to describe the result.

All forces set to trap the extortionists gathered in a group, and in their surprise and disappointment began discussing the queer case in loud tones. A crowd was gathering which was blocking traffic.

"Old Perk" was the first to recover from his surprise.

"Get the hell out of this neighborhood," he yelled to his working forces. "All of you get down to my office!"

The working force dissolved and "Old Perk" drove away.

At "Old Perk's" office shortly afterward a conference of the defeated forces of the law and of science was held.

"Old Perk" stormed and raged and the detective captain in charge fumed and fussed, but nothing came of it all. One was as powerless as another. Finally the conference adjourned.

The next morning in the mail, Perkins Ferguson, president of Schefert Engineering Corporation, received a letter carefully printed in rubber type. It read:

> Thanks for the $50 bill. You cheated us by $99,950. This will never do. Don't be like that. You poor fools, you make us increase our demand. We double it. Leave $200,000 for us on your desk and leave the desk unlocked. We will get it. Every time you ignore one of our demands, one of your number will die. Better take this matter seriously. Last warning.
>
> Invisible Death.

"Not another dime will they get out of me," mused Ferguson.

He started opening the rest of his mail.

A clerk entered and handed him a telegram. It read:

"Damon Farnsworth struck down at breakfast table. Family heard humming sound as he fell from his chair. Removed to Medical Center. Skull reported fractured. May die.

"William Devins, Chief of Police, Larchmont."

Ferguson wildly seized the telephone. "Get me Farnsworth's house at Larchmont!" he shouted to his operator.

The phone was answered by Jones, the butler.

"This is Ferguson."

An agitated voice replied:

"'Ow sir, yes sir. It's true, sir. 'E was bleeding at the 'ead, sir. Something 'it 'im."

"Let me talk to Mrs. Farnsworth."

"They are at the 'ospital, sir."

"One of the boys."

"Both are at the 'ospital, sir."

"Do you think he will live?"

"An' 'ow could I say, sir?"

Ferguson called the Medical Center. They permitted him to talk to a doctor and a nurse. The nurse referred him to the doctor, who said:

"He is unconscious. There is a wicked fracture at the base of the brain. He was struck from the back—a club, I believe. He may die without regaining consciousness. I am hoping he will rally and that he will be all right."

Ferguson ordered his car and, with Lees at his heels, jumped in the tonneau. He heard a humming sound back of him. He looked back and saw nothing. Both he and Lees were too impressed for words.

"Step on it," Ferguson ordered the chauffeur. "Drive us to the Medical Center."

At the world's largest group of hospitals, Ferguson's worst fears were confirmed. The patient was reported sinking.

Ferguson, giant of Wall Street, was a low spirited man as he drove back down town to his office. With Lees he passed through the outer offices, buzzing with business and the click of typewriters. Not a head was raised from a desk or machine. It was a well-drilled force.

Into his private sanctum he walked or rather dragged himself, and wearily he sat down. He pushed a pile of papers from him and ran his hand over his hot brow.

Blood pounded at his temples.

For the first time in his life he faced a situation which was too deep for his understanding.

Over and over again he reviewed the uncanny events as Lees sat awaiting orders.

"I cannot have them killing off my friends like that," he mused finally.

He called a clerk.

"Go to the bank and get $200,000 in fifties and one hundreds," he commanded.

When the clerk returned with the money he laid the package on his desk and left the desk open. "This might appear cowardly, but it will give us time," he said. Lees did not offer an opinion.

Ferguson drew a personal note for $200,000 and sent it to the Schefert Corporation's attorneys. This amount represented a large part of Ferguson's personal assets, not involved with any company with which he was connected. He told Lees to go about his further investigations. Then he left the office and started for his home. "I'll bank my life Lees will have those crooks lined up within a week," he assured himself as he lolled in his auto, bound homeward. But his voice sounded hollow, and the blood still pounded at his temples.

Reaching home, he found a call from the western plant, at Chicago. He phoned the superintendent with a foreboding that all was not well.

"This you, Perk?" sounded the voice on the wire.

"Yes, what's up?"

"I had not intended bothering you with this, but in the light of all that has happened I guess you had better know that one of our engineers went stark mad out here about three weeks ago. He was a very brainy man but his reason snapped. He first appeared queer when he began talking of anarchy and cursing capitalists. Then one afternoon he struck a shop foreman down with a heavy wrench and rushed out of the plant. We have not seen him since. The police have been looking for him, but he is still at large."

"That explains a lot of things," said "Old Perk." "Tell the police to keep after him. We'll look for him here. File me a complete detailed report of the incident by telegraph," he instructed. Then he asked:

"How is the foreman? Badly hurt?"

"He dodged; it was a glancing blow. The foreman was back to work in a week. But he is nervous and has armed himself. We have put on extra guards."

"Good," commended Ferguson. "Don't hesitate to spend tolls to keep me advised of any developments."

An hour and a half later, Ferguson phoned the chief clerk in his offices:

"Go into my private office," he ordered, "and see if there is a package on my desk. It is a bank package."

The clerk returned in a few moments.

"There is no package on your desk, Mr. Ferguson."

"That is all I wanted to know," said Ferguson, and hung up the receiver.

Then Ferguson called up the Darrow home and tried to get in touch with Lees, but was unable to do so, as Mrs. Darrow said she had not seen him since he had been called back to the office.

The reason Ferguson could not reach Lees was because Lees had decided to learn once and for all if Jouret wore number nine shoes. He had started for Jouret's in his own car. It was a beautiful country he was traversing, but he had no time to note that the tree branches almost met over his head and that his way was bordered with a profusion of wild flowers, displaying a rainbow of colors.

The house of Jouret, the retired circus performer, sat back far from the road, against the side of a beautiful hill, and was surrounded by poplars. The landscape was wilder and more natural than that of the Darrow place adjoining.

The door was opened by a Porto Rican boy. Lees lost no time. He said bluntly:

"Tell your master that a gentleman is here to see him on very particular business."

Jouret, himself, came back with the boy.

"What is it?" he asked, smiling a welcome.

"I am working on the case of the death of Mr. Darrow, your neighbor. I believed you might have seen something. I thought you might aid me."

Jouret betrayed no surprise.

"Come in," he said. He led the way to a large reception room and asked his visitor to be seated. He was the soul of affability. Short, husky and florid.

His eyes large, black and staring. His hair black, quite long and curling upward at the ears. He was dressed in black, and he had the appearance of a big, fat crow.

"I am glad you came," he greeted his guest, "for I have far too few callers." He switched on a big electric bunch-light in the center of the room, for it was dusk.

"We have been told that you are a retired circus man," said Lees, in his usual frank manner.

"Not exactly," said Jouret. "I traveled on the continent, finally journeying to Australia and then to the States. I crossed the country from San Francisco and settled down here. I was known as 'Elias, the Great.' I had my own company and property. It was a magic show. It was not a circus, although we did carry two elephants, three camels, some ponies, snakes, and birds and smaller animals. That's where the circus report came from.

"When I retired I sold my stock to a circus. The newspapers regarded it as funny, and one of them printed a half page story with pictures about the public sale. It was very much exaggerated. They mentioned giraffes, hyenas, and a lot of other animals I never possessed. Odd, wasn't it, getting so much publicity after I was through needing it? However I never, in those days, dodged the limelight." Jouret ended his speech with a loud and hearty guffaw.

"I will call my daughter," Jouret appended. "She will be glad to meet you." He left the room.

Lees had taken occasion to note the size of Jouret's feet. They were small, almost effeminate. More likely fives or sixes than nines.

Soon Jouret returned with a girl in her early twenties. She was blond and radiantly beautiful.

Doris Jouret bowed and smiled in a perfectly friendly manner. Lees noted that there was something about her eyes that made her appear dazed.

Jouret monopolized the conversation, giving no one a chance to edge in a word.

"This gentleman desires information in connection with the death of our neighbor Mr., or is it Dr., Darrow? I want you to assure him, as I will, that we have seen or noted nothing that could possibly throw light on the strange case."

The girl nodded, it seemed a little wearily, and Jouret was off on another conversational flight:

"I too am a man of scientific attainments," he chattered. "I am a biologist, toxicologist, doctor of medicine, a geologist, metalurgist, mineralogist, and somewhat of a mechanic and electrician. I have given long hours to the study of strange sciences in meta-physics, to which you men give too little attention. There are sciences which transcend any of this sphere. There is a higher astronomy. I neglected to say that I am an astronomer."

"Yes?" drawled Lees.

"Yes!" said Jouret emphatically.

The girl had adopted rather a theatrical pose, which disclosed considerable of her nether charms, and said nothing at all.

"When you find your man," volunteered Jouret, "you will find a madman." He said this ponderously and with a gesture meant evidently to be impressive.

"You believe a madman did it?" asked Lees, as Jouret paused, expecting a question.

"Undoubtedly. It was a paranoic with delusions of money, grandeur and a strongly developed homicidal mania. To me, that is the only sensible solution. I am quite sure that I am correct."

Lees arose to go and Jouret did not urge him to stay. He bowed Lees out and Doris bowed with him.

"She is a beautiful girl," mused Lees once he was outside.

Lees ran over in his mind the circumstances of his visit to Jouret. There was no doubt in his mind that Jouret's shoes were too small to be number nines, and he reasoned that that fact might tend to eliminate Jouret. But he was not satisfied.

"I am going to get some gas," he told himself, "and then I am going to get two private detectives to assist me, for I'm going right back there. For the first time in my life I am going to be a Peeping Tom.

"There is no moon. The poplars will give us a view of all three floors of that house, if they leave their blinds up enough, and three of us can watch all three floors at once."

He phoned Ferguson that he might be busy for days, joined his pair of operatives from the detective agency and for some time the three operated on a well conceived plan.

It was probably a week later that Lees rendered a report to Perkins Ferguson, which for a time proved one of the strangest documents in the weird case. It read:

"You will probably think I am crazy, and for this reason I am having this report subscribed and sworn to, jointly and severally. With my two detectives I have seen Miss Jouret, the girl I told you about over the phone, in three places at one and the same time. Not once but twice this has happened.

"Looking through the windows of the Jouret place at night, we saw the girl on the first, second and third floor of the house. We believed this due to a clever arrangement of mirrors. But figure this out:

"The next day she drove a car to town. We followed. She got out at one theater and entered. She did not come back, that we could see, but the car drove off. There was no chauffeur, and we thought we had discovered the driverless auto, until we looked and saw Miss Jouret still at the wheel.

"She got out and entered another theater. She did not come back, but the car drove off with her still at the wheel. She entered a third theater after parking the car and this time the driver's seat and the tonneau was empty.

"Reverse the reel and you will see her coming out of three theaters and driving home. That is what happened. There must be three of her, all identical, but only one shows at a time. If it's some of Jouret's far-famed magic, I'll say he's some conjurer. The explanation is not yet forthcoming. We want to shadow Jouret, but he never goes anywhere. The girl has only been out the one time when she attended three matinees as described. Believe it or not.

"The next night we each—the two detectives and I—tried to steal a march on one another and called her up and asked her to go out. To our individual surprise, she agreed in each case. To our collective surprise, she kept all three dates on the same night. She walked through the trees in this vicinity with me. She also drove down the road in the auto with one of my detectives, and she went dancing with the other. She was in three places miles apart at one and the same time.

"We each brought her home within a half hour of the other and we are swearing to that. Either we are all hypnotized or else there are three identical Misses Jouret.

"Jouret himself treats us all wonderfully, gives us the run of the house, and tries to talk us to death."

The strange document was subscribed by Lees and the two detectives and was held by Ferguson pending developments.

The next report from Lees read:

"I had a chance to prowl around the Jouret house a little while waiting for Miss Jouret to dress. I met her twice in my ramblings and a few minutes later she met me again, this time in a different costume.

"I got a chance to search the woods back of Jouret's house in the evening. I found a spot where the earth had been disturbed, and dug up a pair of shoes. They were number nines."

A fourth report from him read:

"We found the body of the crazed engineer. He had drowned himself in a lake. This eliminates him as a murder suspect."

Two weeks passed with no new developments in the "Invisible Death" case except for the arrival of a letter demanding $1,000,000 and threatening the life of Perkins Ferguson if the demand was ignored. It was ignored, and only served to spur Lees and his detectives on to decisive action.

They decided to rush the Jouret house and kidnap Jouret with the idea of holding him until he agreed to explain the presence of the number nine shoes buried back of his house.

A low moon hung over the poplars when Lees rang the Jouret front door bell. One detective was guarding a side door and the other a back door.

Suddenly Jouret was seen to jump from a second-story window. As he did, a car driven by one of his Porto Ricans came along the drive and he leaped into it. Lees, first to see Jouret, called his detectives. They came running. Their car was waiting in the road.

The Porto Rican was seen to jump from the Jouret car just as it started south towards New York.

Lees took up the race. Both cars had plenty of power, but the Jouret car suddenly disappeared as a low humming noise began to break the stillness of the night.

One of the detectives was at the wheel. Lees, as usual, was giving orders:

"Keep close to that hum. Never mind that you cannot *see* the car. It is there all right. If you can gain on it enough, drive right into it."

"Righto!" shouted the detective. "We're wise to him now."

The humming noise was taking on speed with every second. So was Lees' car. Soon Lees' car was making sixty miles an hour with the hum just ahead and barely audible.

Past traffic lights, over bridges and grade crossings the mad chase of the phantom continued.

Wildly racing through the night, missing other cars by a breath, the big, visible auto continued its pursuit of—what?

Careening, Lees' car rounded a curve, and, above the hum just ahead, they heard the shouted curses of their quarry. But he could not be seen. Lees could only see the road marked by his lights.

Mile after mile the wild, uncanny chase of the phantom continued.

Soon the lights of New York could be seen in the distance. The cars were forced to slow down somewhat. Suddenly there was a thundering crash ahead. A car was twisted in a mass of tangled wreckage.

Feminine and masculine shrieks blended as Lees' car piled up on the wrecked heap. A third car, becoming suddenly visible, rolled over and brought up at the edge of the road. From this car emerged the limping, cursing form of Jouret.

From the wreckage three painfully injured young men dragged and tore themselves. Then they leaped—ignoring their hurts—at the limping figure.

The fight was on. Jouret was heavy and powerful and proved an obstinate fighter, for he knew he was fighting for his life. He bit and clawed. He kicked with one uninjured leg and butted with his massive head.

Lees and his detectives were fighting with no respect for the rules. Lees managed to get his two hands on the bull-neck of Jouret just as one detective connected a duet of blows to the man's wind.

Lees' hands closed in a steely grip, and soon Jouret was limp and helpless.

They held him there. An ambulance arrived. A few minutes later a police auto with reserves came on the scene. The police shackled Jouret.

The car that had been hit by the phantom was a light sedan. It was occupied by two women. Their bodies were drawn from the wreckage. Both were dead—innocents sacrificed to the blood madness of a maniac.

Jouret was right about himself. He was a paranoic with a strongly developed homicidal mania.

In the wreckage was found a package containing $200,000 and also two twisted and broken mechanisms. One of these was about the size of an ordinary kitchen coffee-mill, and the other slightly larger.

Regarding these machines, Lees wrote in a report:

"While making a fourth search of Darrow's laboratory, I found the equations, specifications and what I believe to be the full plans for the last invention of the ingenious Darius Darrow.

"Many of the most astounding inventions and discoveries have resulted from theories which were laughed to scorn at the time they were advanced. Roebling's plans for the Brooklyn Bridge resulted in a meeting of the foremost engineers of the day. All agreed that the plans were built on a false premise. They argued that the bridge would fall of its own weight. Then they all had a good laugh. The bridge still stands.

"Watching smoke float over a hill from army camp fires caused an early French scientist to dream of filling a bag full of smoke and riding with it over the hill. The first balloon was the answer to this dream.

"James Watt is said to have gotten his idea for a steam engine from watching a lid on a tea-kettle dance under steam pressure.

"When Langley was flying his man-carrying kites the Wright brothers dreamed of hitching an engine and a propeller to a giant kite. The airplane was the result of these experiments.

"Darrow got his idea from watching a rapidly revolving wheel. He noticed that the spokes and rim blended into a blurred disc when a certain speed was reached. The entire wheel was practically invisible, under certain lighting conditions, when a higher speed was attained.

"Darrow went further and reached the conclusion that there was a rate of vibration that would produce invisibility. This was accepted in practically all engineering research plants, long before it was perfected by Darrow.

"The facts are that any rapidly vibrating object becomes more and more difficult to outline as its rate of vibration increases. All that was left for Darrow was to arrive at the exact mathematical time, tone, or rate of vibration producing invisibility and to construct a vibrator tuned to produce this condition.

"His first machine produced the vibrations of invisibility in a field with a three-foot radius in all directions. That is, it caused every solid object, within this atmospheric field, to vibrate at the rate, tone, or speed of invisibility. This machine was in no sense rotary. It departed from the original example of a revolving wheel and entered instead into general vibration in a given or measured field.

"The pulsations or vibrations of an ordinary automobile engine will cause every ounce of metal, or solid, in the automobile—including the driver—to vibrate at the same rate or momentum. This is a known fact, and it provided the basis for Darrow's experiments.

"Darrow built two machines. The first had a field with a radius of three feet on all sides. This was used by the killer in his murders. Jouret stole this machine first, thus paving his way for the second robbery.

"With the first machine in his possession, Jouret was able to commit the Darrow murder without being seen. He had to have the second and larger machine, however, to make his auto disappear. He stole the larger machine at the time of the Darrow murder, and with it he had his auto vanish, as the gardener testified.

"Both machines were hopelessly smashed in the wreck, but with Darrow's documents at hand, we might be able to construct another and a larger model. A machine built on the proper scale will make a plane or a battleship invisible and should, as Darrow said, make war against this country impossible.

"Digging into Jouret's history we found that the 'Misses Jouret' were one-cell triplets. Their mother, Mrs. Doris Nettleton, an English woman, was a member of Jouret's troupe, as was the father.

"The mother died at the birth of the triplets. The father died a few years later. The company was touring Australia at the time. Jouret and the father had the birth of only one baby recorded. She was named Doris, after the mother. The other girls also used this one name. They now have only one name among them until the court gives them individual names.

"Jouret never let but one girl be seen at a time. The reason was that he and the father had planned to use the girls, when grown, to create a surprising stage illusion. In this illusion, one girl was to act as the earthly body and the other girls as the astral bodies of the same purported individual.

"The father died, and Jouret retired before he ever got around to staging the illusion. Jouret continued the deception, however, because it appealed to his showman's nature.

"The girls, at all times, were under the hypnotic control of Jouret, and, of course, knew nothing of his crazed intellect or crimes. Upon his arrest Jouret released the girls from the spell of years.

"The Misses Nettleton say that Jouret was always kind to them and was an ethical showman until his mind gave way.

"I told the triplets that I might find them employment with our concern, but they prefer to follow in the footsteps of their mother and father, and return to the stage."

Ferguson, quite his normal self once more, since Farnsworth was recovering slowly, twitted Lees about being in love with one of the triplets. Lees admitted they were most gorgeous blondes, but insisted he preferred one brunette.

"Then another thing," added Lees. "Any man who falls in love with one of the Nettleton triplets will never be sure just which one he fell in love with."